Table of Contents

Fire Service Instructor

Fifth Edition

VALIDATED BY

INTERNATIONAL FIRE SERVICE TRAINING ASSOCIATION

PUBLISHED BY

FIRE PROTECTION PUBLICATIONS • OKLAHOMA STATE UNIVERSITY

Cover Photo Courtesy of: Columbia, Missouri Fire Department

Dedication

This manual is dedicated to the members of that unselfish organization

of men and women who hold devotion to duty

above personal risk, who count on sincerity of service above

personal comfort and convenience, who strive unceasingly to find

better ways of protecting the lives, homes and property

of their fellow citizens from the ravages of fire and other

disasters . . . **The Firefighters of All Nations.**

Dear Firefighter:

The International Fire Service Training Association (IFSTA) is an organization that exists for the purpose of serving firefighters' training needs. Fire Protection Publications is the publisher of IFSTA materials. Fire Protection Publications staff members participate in the National Fire Protection Association, International Association of Fire Chiefs, and the International Society of Fire Service Instructors.

If you need additional information concerning our organization or assistance with manual orders, contact:

Customer Services
Fire Protection Publications
Oklahoma State University
Stillwater, OK 74078-0118
1-(800) 654-4055

For assistance with training materials, recommended material for inclusion in a manual or questions on manual content, contact:

Technical Services
Fire Protection Publications
Oklahoma State University
Stillwater, OK 74078-0118
(405) 744-5723

First Printing, July 1990
Second Printing, June 1991
Third Printing, May 1994

LIST OF TABLES

THE INTERNATIONAL FIRE SERVICE TRAINING ASSOCIATION

The International Fire Service Training Association (IFSTA) was established as a "nonprofit educational association of fire fighting personnel who are dedicated to upgrading fire fighting techniques and safety through training."

This training association was formed in November 1934, when the Western Actuarial Bureau sponsored a conference in Kansas City, Missouri. The meeting was held to determine how all the agencies interested in publishing fire service training material could coordinate their efforts. Four states were represented at this initial conference. It was decided that, since the representatives from Oklahoma had done some pioneering in fire training manual development, other interested states should join forces with them. This merger made it possible to develop training materials broader in scope than those published by individual agencies. This merger further made possible a reduction in publication costs, since it enabled each state or agency to benefit from the economy of relatively large printing orders. These savings would not be possible if each individual state or department developed and published its own training material.

To carry out the mission of IFSTA, Fire Protection Publications was established as an entity of Oklahoma State University. Fire Protection Publications' primary function is to publish and disseminate training texts as proposed and validated by IFSTA. Secondary, Fire Protection Publications researches, acquires, produces, and markets high-quality learning and teaching aids as consistent with IFSTA's mission. The IFSTA Executive Director is officed at Fire Protection Publications.

IFSTA's purpose is: to validate training materials for publication; to develop training materials for publication; to check proposed rough drafts for errors; to add new techniques and developments; and to delete obsolete and outmoded methods. This work is carried out at the annual Validation Conference.

The IFSTA Validation Conference is held the second full week in July, at Oklahoma State University or in the vicinity. Fire Protection Publications, the IFSTA publisher, establishes the revision schedule for manuals and introduces new manuscripts. Manual committee members are selected for technical input by Fire Protection Publications and the IFSTA Executive Secretary. Committees meet and work at the conference addressing the current standards of the National Fire Protection Association and other standard-making groups as applicable.

Most of the committee members are affiliated with other international fire protection organizations. The Validation Conference brings together individuals from several related and allied fields, such as:

...key fire department executives and training officers,

...educators from colleges and universities.

...representatives from governmental agencies,

...delegates of firefighter associations and industrial organizations, and

...engineers from the fire insurance industry.

Committee members are not paid, nor are their expenses covered by IFSTA or Fire Protection Publications. They come because of commitment to the fire service and its future through training. Being on a committee is prestigious in the fire service community, and committee members are acknowledged leaders in their fields.

This unique feature provides a close relationship between the International Fire Service Training Association and other fire protection agencies, which helps to correlate the efforts of all concerned.

IFSTA manuals are now the official teaching texts of most of the states and provinces of North America. Additionally, numerous U.S. and Canadian government agencies as well as other English-speaking countries have officially accepted the IFSTA manuals.

Preface

There continues to be a need for competent and dedicated fire service instructors. Fire protection costs are skyrocketing, fire control methods are changing, and on-duty personnel are decreasing. Now, more than ever, firefighters must be trained to make the best use of available resources and perform job skills efficiently.

This edition of **Fire Service Instructor** includes the information and techniques necessary to meet the challenges facing fire service instructors. The material has been updated and extensively revised to help the firefighter meet the requirements of NFPA 1041, *Fire Service Instructor Professional Qualifications,* Levels I and II.

Fire Service Instructor was compiled and validated by a committee of the International Fire Service Training Association. Acknowledgement and grateful thanks are offered to the continuing committee members who provided input and worked so diligently updating and revising existing material.

Chairman
Bill Vandevort
Fire Service Training Specialist
State Fire Marshal's Office
Sacramento, California

Vice-Chairman
Gerald E. Monigold
Director, Fire Service Institute
University of Illinois
Champaign, Illinois

The editorial staff also extends its appreciation to the other dedicated committee members who devoted their time and talents toward this manual.

Wesley Beitl	Sam Goldwater
Springfield, Illinois	Angier, North Carolina
Ron Hamm	Randy Hood
Ft. Wayne, Indiana	Shreveport, Louisiana
Richard Resurreccion	John Ryan
Long Beach, California	Las Vegas, Nevada
Lloyd Scholer	Chuck Wilson
Millville, Minnesota	Vista, California

Special thanks are due to Ron Payne of the Audio-Visual Center at Oklahoma State University for his assistance on the Training Aids chapter. Special thanks are also due to J.R. Bachtler for supplying information for the Safety chapter.

We are grateful to James E. (Doc) Daugherty and Training Officer Gary Warren of the Columbia (Missouri) Fire Department and the members of that organization for the assistance with photographs in this publication.

The exceptional artwork in the manual was done by our own Lori Schoonover. Her talent provides a very special effect and it is greatly appreciated.

Gratitude is also extended to the following individuals whose contributions made the final publication of this manual possible:

William Westhoff, Senior Publications Editor
Lynne C. Murnane, Senior Publications Editor
Carol Smith, Associate Editor
Cynthia Brakhage, Publications Specialist
Robert Fleischner, Publications Specialist
Beth Ann Chlouber, Publications Specialist
Susan Walker, Instructional Development Coordinator
Don Davis, Production Coordinator
Ann Moffat, Senior Graphic Designer
Desa Porter, Graphic Designer
Karen Murphy, Phototypesetting Technician
John Hoss, Research Technician
Terri Jo Gaines, Senior Clerk/Typist

Gene P. Carlson
Assistant Director
Fire Protection Publicaitons

COPYRIGHT LAW
AN EDUCATOR'S RESPONSIBILITIES

The law limits what you may copy, under what conditions you may copy, and for what purpose you may copy. Authors and publishers have specific rights under the law. However, the law permits educators access to information and to copy materials under clearly defined guidelines.

These guidelines include:

- The purpose and character of use: whether the purpose is commercial or nonprofit educational

- The nature of the copyrighted work (example: Is the work a textbook meant for classroom use?)

- The amount and substantiality copied in relation to the copyrighted work as a whole

- Whether the copied material will affect the potential sales or value of the copyrighted work

A single copy may be made by a teacher, by request, for research or teaching of a chapter from a book; an article from a periodical; short stories or essays; or a chart, diagram, or drawing. Multiple copies for the classroom must not exceed one copy per student and must meet the test of "brevity, spontaneity, and cumulative effect," which are defined as follows:

- Brevity is either a complete article, story, or essay of less than 2,500 words or an excerpt from any work of not more than 1,000 words or 10 percent of the work, whichever is less for example, (one chart, graph, diagram, drawing, cartoon, or picture per book or periodical issue).

- Spontaneity is (if permission is not sought prior to use) the decision to use the article or excerpt so close in time that it would be unreasonable to expect a timely reply to a request for permission.

- Cumulative Effect is no more than one piece copied from the same author nor no more than three articles from the same periodical volume during one classroom term. There can be no more than nine instances of multiple copying per course, per class term.

Under no circumstances can there be copying of or from works intended to be consumable in the course. These include workbooks, answer sheets, and the like. Copying shall not substitute for the purchase of books, publisher's reprints, or periodicals.

The copyright law specifies a monetary penalty for legal damages for each violation. Even a defendant (individual and/or the organization) not found in violation must bear court costs and attorney's fees.

Permission to copy is obtained by writing to the publisher with the following information:

- Title, author(s), and editor(s)
- Edition and/or issue
- Exact amount of material to be copied
- Nature of use, including if it is for resale
- How material will be reproduced
- Number of copies to be made

Copies of copyrighted materials must include a credit line to the original work, author or editor, and the publisher with copyright notice or reprint permission.

Glossary

A

ACCIDENT — An unplanned, uncontrolled event that results from unsafe acts of persons and/or unsafe occupational conditions, either of which can result in injury.

AFFECTIVE LEARNING — Learning that relates to interests, attitudes, and values.

AFFIRMATIVE ACTION — Employment programs designed to make a special effort to identify, hire, and promote special populations where the current labor force in a jurisdiction or labor market is not representative of the overall population.

APPLICATION STEP — The third step in the four-step teaching method of conducting a lesson, in which the learner is given the opportunity to apply what has been learned and to perform under supervision and assistance as necessary.

B

BEHAVIORAL OBJECTIVE — A measurable statement of behavior required to demonstrate that learning has occurred.

BLOCK — A division of an occupation consisting of a group of related tasks with some one factor in common.

C

CENTRAL PROCESSING UNIT (CPU) — The part of a computer that actually processes information.

COGNITIVE LEARNING — Learning that relates to knowledge and intellectual skills.

COMPETENCY-BASED LEARNING (CBL) — Training based upon the competencies of a profession or job. Competencies are the absolute standards or criteria of performance. Emphasis is on what the learner will learn. (Same as criterion-referenced and performance-based.)

COMPUTER-AIDED INSTRUCTION (CAI) — An instructional approach that uses the computer to present instruction to the student on an individualized, self-paced basis.

COPYRIGHT LAW — A law designed to protect the competitive advantage developed by an individual or organization as a result of their creativity.

COURSE DESCRIPTION — Relates the basic goals and objectives of the course in a broad, general manner. It is designed to provide a framework and guide for the further development of the course and also communicate the course content.

COURSE OBJECTIVES — A specific identification of the planned results of a course of instruction.

COURSE OUTLINE — A list of jobs and information to be taught to fulfill previously identified needs and objectives.

CRITERIA — One of the three requirements of evaluation. The standard against which student learning is compared after instruction. The expected learning outcome. Examples are behavioral objectives or NFPA standards.

CRITERION-REFERENCED TESTING — The measurement of individual performance against a set standard or criteria, not against other students. Mastery learning is the key element to criterion-referenced testing.

D

DEMONSTRATION — A teaching method that includes the act of showing a person how to do something.

DESKTOP PUBLISHING — Using a computer to develop manuals, handouts, and so on.

DISCUSSION — A teaching method where students contribute to the class session by using their knowledge and experience to provide input.

DOMAINS OF LEARNING — Areas of learning and classification of learning objectives, which are often referred to as cognitive (knowledge), affective (attitude), and psychomotor (skill) learning.

E

EQUAL EMPLOYMENT OPPORTUNITY — A personnel management responsibility to be sensitive to the social, economic, and political needs of a jurisdiction or labor market.

EVALUATION — The systematic and thoughtful collection of information for decision making. It consists of criteria, evidence, and judgment.

EVALUATION STEP — The fourth step in conducting a lesson, in which the student demonstrates that the required degree of proficiency has been achieved.

EVIDENCE — One of three requirements of evaluation. The information, data, or observation that allow the instructor to compare what was expected to what actually occurred.

F

FAMILY EDUCATION RIGHTS AND PRIVACY ACT OF 1974 — Provides that an individual's records are confidential and that information contained in those records may not be released without the individual's prior written consent.

FEEDBACK — Student responses, generated by questions, discussions, or opportunities to perform, that demonstrate learning or understanding.

FORESEEABILITY — The concept that instruction should be based not only on dangerous conditions that may exist in training, but also anticipate what firefighters might face on the job.

FORMATIVE EVALUATION — The ongoing, repeated assessment during course development and during or after instruction to determine the most effective instructional content, methods, aids, and testing techniques.

FOUR-STEP TEACHING METHOD — A teaching method based upon four steps: preparation, presentation, application, and evaluation.

G

GRADING SYSTEM — The system used to convert achievement to grades or class standing.

H

HARDWARE — Refers to the computer, electronic components, keyboard, CRT, disk drives, and other physical items connected with the computer system.

I

ILLUSTRATION — Method of teaching that uses the sense of sight. Showing by illustration includes the use of drawings, pictures, slides, transparencies, film, models, and other visual aids that may clarify details or processes.

INDICATOR ACTION — The part of a behavioral objective that tells how a student will show a desired behavior so it can be observed and measured.

INDIVIDUALIZED INSTRUCTION — The process of matching instructional methods and media with learning objectives and students' learning styles.

INFORMATION PRESENTATION — A lesson plan format that addresses the cognitive objectives by introducing new information, facts, principles, and theories.

INFORMATION SHEET — An instructional sheet used to present ideas or information to the learner; used when desired information is not in printed form or otherwise available to the student.

INSTRUCTION ORDER — The organization of jobs or ideas according to learning difficulty so that learning proceeds from the simple to the complex.

INSTRUCTIONAL DEVELOPMENT PROCESS — The process of designing classroom instruction that consists of three major components: analysis, design, and evaluation.

INSTRUCTOR — The person charged with the responsibility to conduct the class, direct the instructional process, teach skills, impart new information, lead discussions, and cause learning to take place.

J

JOB — An organized segment of instruction designed to develop psychomotor skills or technical knowledge.

JOB BREAKDOWN SHEET — An instructional sheet listing step-by-step procedures and required knowledge. It is designed to assist in teaching and learning a psychomotor objective.

JUDGMENT — One of three requirements of evaluation. The decision-making ability of the instructor to make comparisons, discernments, or conclusions about the instructional process and learner outcomes.

K

KEY POINTS — Factors that condition or influence operations with an occupation. Information that must be known to perform the operations in a job.

L

LAW OF ASSOCIATION — The principle that learning comes easier when new information is related to similar things already known.

LAW OF EFFECT — The notion that learning is more effective when a feeling of satisfaction, pleasantness, or reward accompanies or is a result of the learning process.

LAW OF EXERCISE — The idea that repetition is necessary for the proficient development of a mental or physical skill.

LAW OF INTENSITY — The premise that if the experience is real, there is more likely to be a change in behavior or learning.

LAW OF READINESS — The concept that a person learns when physically and mentally adjusted or ready to receive instruction.

LAW OF RECENCY — The principle that the more recently the reviews, warm-ups, and make-up exercises are practiced just before using the skill, the more effective the performance will be.

LEARNING — A relatively permanent change in behavior that occurs as a result of acquiring new information and putting it to use through practice.

LECTURE — A teaching method in which the instructor verbally relays information to teach a lesson.

L-E-A-S-T METHOD — A progressive discipline method used in the classroom.

LESSON PLAN — An outlined plan for teaching, listing pertinent teaching information and using the four-step teaching method.

LEVEL OF LEARNING (INSTRUCTION) — The depth of instruction for a specific skill and/or technical information that enables the student to meet the minimal requirements of the occupation.

LIABILITY — A broad, comprehensive term describing a person or organization's responsibility in the law. This responsibility implies that if a wrong has occurred, a person or an organization must respond to legal allegations.

M

MANIPULATIVE LESSON — Another term for the practical demonstration portion of a lesson plan.

MANIPULATIVE-PERFORMANCE TEST — Practical competency-based tests that measure mastery of the psychomotor objectives as they are performed in a job or evolution.

MANIPULATIVE SKILLS — Skills that use the psychomotor domain of learning. Refers to the ability to physically manipulate an object or move the body to accomplish a task.

MASTERY LEARNING — An element of criterion-referenced or competency-based learning. Outcomes of learning are expressed in minimum levels of performance for each competency.

MATERIALS NEEDED — A list of everything needed to teach a lesson: models, mock-ups, visual aids, equipment, handouts, quizzes, and so on.

METHOD OF INSTRUCTION — A procedure, technique, or manner of instructing others, determined by the type of learning to take place. Typical examples are lecture, demonstration, or group discussion.

MINIMUM ACCEPTABLE STANDARD — The lowest acceptable level of student performance.

MOTIVATION — An internal process in which energy is produced by needs or expended in the direction of goals.

N

NORM-REFERENCED TESTING — The measurement of student performance against other students, with an emphasis on discriminating among students and assigning grades.

O

OCCUPATIONAL ANALYSIS — A method to gather information about an occupation, develop a description of qualifications, conditions for performance, and an orderly list of duties.

OPERATION — One step in performing a job skill within an occupation. Operations are listed in the order in which they are performed.

P

PERFORMANCE LEVELS — The desired level of ability in doing a particular job as specified in a student-performance goal.

PERFORMANCE STANDARDS — The benchmarks for judging progress toward goals and objectives.

PERFORMANCE TESTS — Tests given in the middle or at the end of the instruction to measure final performance.

PERSONAL COMPUTER (PC) — A computer designed to be used as a single workstation.

PLAGIARISM — To present as an original idea without crediting the source.

PRACTICAL DEMONSTRATION — That part of the lesson plan that addresses psychomotor objectives with a prepared job breakdown sheet.

PREPARATION STEP — The first step in conducting a lesson, in which the job or topic to be taught is identified, a teaching base is developed, and students are motivated to learn.

PRESCRIPTIVE TESTS — Tests given at the beginning of instruction to determine readiness to learn or to determine what the student already knows.

PRESCRIPTIVE TRAINING — An instructional approach that uses the four-step teaching method in a different order; evaluation, preparation, presentation, application, and re-evaluation.

PRESENTATION STEP — The second step in conducting a lesson, in which new information and skills are presented to the learners.

PRODUCTION ORDER — The order in which jobs must be done. The more difficult jobs may have to be done first.

PROGRESS CHART — A chart designed to record the progress of an individual or group during a course of study.

PROGRESS TESTS — Tests given during instruction to measure improvement and to diagnose learning difficulties.

PROXIMATE CAUSE — One that in a naturally continuous sequence produces the injury, and without which, the result would not have occurred.

PSYCHOMOTOR LEARNING — Learning that relates to physical or manipulative skills.

R

RELIABILITY — The consistency and accuracy of measurement in a test. A condition of validity.

RIGHT OF PRIVACY — The concept that means that an individual's records are confidential. See Family Education Rights and Privacy Act of 1974.

S

SESSION GUIDE — A plan for using a group of lesson plans or instructional materials during a predetermined period of instructional time.

SOFTWARE — A computer program that performs a specific function or set of functions.

STANDARD DEVIATION — The average of the degree to which the scores in a test deviate from the mean.

STUDENT — The most important member of any class in that all the activities and efforts are directed toward enabling him or her to learn.

STUDY SHEET — An instructional sheet designed to arouse interest in the assignment and to provide instructions for additional or outside study by the student.

SUMMATIVE EVALUATION — A comprehensive approach to evaluation, using test results, instructor observations, and the course critique to determine the total course effectiveness.

T

TASK — A combination of duties and jobs in an occupation that are performed regularly and require psychomotor skills and technical information to meet occupational requirements.

TECHNICAL LESSON — Another term for the information presentation portion of a lesson plan.

TECHNICAL SKILLS — Skills involving manipulative aptitude.

TEST ITEM ANALYSIS — A process that shows how difficult a test is, how much it discriminates between high and low scorers, and whether the alternative used for distracters truly work.

TEST PLANNING — The planning steps to determine the purpose and type of test, identify and define the learning objectives, prepare the test specifications, and construct the test items.

TORT — A civil wrong or injury.

TRAINING AIDS — Teaching aids such as films, pictures, charts, maps, drawings, and posters.

TRANSFER OF LEARNING — The opportunity to apply what has been learned in one situation to a new situation.

U

UNIT — A division of a block within an occupation consisting of an organized grouping of tasks within that block.

V

VALIDITY — The extent to which a test measures what it says it is to measure.

W

WORD PROCESSOR — A software program used with a personal computer, designed specifically for creating text documents.

Introduction

During the past few years, there have been many changes in fire service training. For many years, fire service training was concerned mostly with a few basic manipulative skills and was conducted on a department-by-department basis. Now, training is often conducted through multi-department drills, regional academies serving several departments, and fire science programs in community colleges.

The expansion of fire service training has necessitated the establishment of a large corps of fire department personnel as instructors. Like all other instructors, fire service instructors must be trained to teach. The qualities a person must have to become a good instructor include the ability to get along with people, a willingness to do the necessary preparation work, and the desire to teach. Since fire service training makes great use of vocational education methods, instructors must acquire knowledge and training in the use of those methods.

Training is the key to fire department efficiency, and fire service instructors must be recognized as key personnel within a fire service organization. Even though specific individuals may be assigned the responsibility of planning and presenting instruction to department members, it does not relieve other officers of their responsibility to participate in the department training program. Providing an effective training program is an important responsibility that requires a commitment from all officers in the department.

With the advent of the NFPA 1041, *Fire Service Instructor Professional Qualifications,* performance standards have been identified for fire department instructors. This standard specifies the information instructors need to meet minimum knowledge and skill level requirements. It should be the goal of every fire department that all current and potential fire service instructors meet the minimum qualifications of the standard. By doing so, fire service personnel interested in teaching will be better prepared to present instruction to their fellow firefighters. Knowledge of proper teaching techniques will result in better instruction and improved efficiency in the performance of job skills.

PURPOSE AND SCOPE

Since training is an essential factor in efficient fire fighting and is dependent upon well-prepared instructors, fire service instructor training is an important responsibility of a fire department. The purpose of this manual is to make information available from which an effective instructor training program can be developed.

The material contained in **Fire Service Instructor** also provides the information called for by NFPA 1041 to meet the minimum knowledge and skill levels for Fire Instructor I and II.

This manual includes an overview of the instructor and the job, factors that influence teaching and learning, and the techniques of preparing and presenting an effective lesson. A chapter is devoted to a discussion of a wide variety of training aids used in instruction. Since many training programs operate on a limited budget, an appendix deals with training aids that can be developed by the instructor. Another chapter discusses the purpose and principles of testing and evaluation. Types of tests and test construction are also covered. **Fire Service Instructor** provides the knowledge base that a fire service instructor needs to effectively prepare and present instruction. An entire appendix is devoted to a model instructional package. Today's fire service instructors are developing the leaders of tomorrow.

Instructional Challenges of the 90s

This chapter provides information that addresses performance objectives in NFPA 1401, *Fire Service Instructor Professional Qualifications* (1987), particularly those referenced in the following sections:

NFPA 1401

Fire Service Instructor

3-2.1

3-2.2

3-3.1

3-3.2

Chapter 1
Instructional Challenges of the 90s

From the time individuals enter school, teachers play an important role in shaping values, beliefs, and attitudes. Teachers shape individuals' views of themselves and the world. Looking back, everyone can probably remember a teacher who was influential in their lives.

As teachers, fire service instructors have the opportunity to positively influence recruits as well as all levels of the organization through leadership and example. Instructors are leaders in the fire service, and fire service personnel can benefit from their experience and knowledge (Figure 1.1). Fire service instructors

Figure 1.1 Instructors are in leadership positions and can use their knowledge and experience to help students learn.

have the clear-cut responsibility to teach the technical aspects of the fire service. Less obvious but equally as important, instructors are also responsible for promoting positive values and motivation to recruits as well as to others in the department. The initial introduction to the fire service will greatly influence a person's attitude toward his or her involvement in a fire service career.

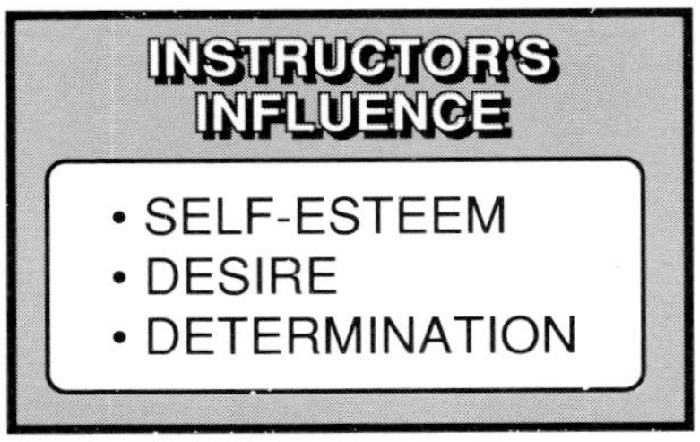

Fire service instructors can have a positive effect on a person's self-esteem, desire to learn, and determination to succeed (Figure 1.2). All seasoned firefighters can remember fire service instructors who have had an impact on their career. They may have motivated the firefighter to make a career of the fire service or given an individual the confidence to handle the job of fire fighting. These instructors are difficult to forget and have a lasting influence in the mind of the learner. Many firefighters are grateful to their previous instructors. The instructors, in turn, are satisfied with the knowledge that they have helped the fire service by positively influencing individuals in the organization.

THE ROLE OF THE INSTRUCTOR

The fire service instructor's primary role is to plan and conduct training. The instructor is responsible for ensuring that performance standards are met and can be measured effectively. Feedback and evaluation are important aspects of the training program.

The fire service has performance standards that must be used to evaluate training. The National Fire Protection Association Standards detail minimum qualifications for fire service personnel. The instructor has to do more than just run the recruits through the training program. An effective program dictates that the instructor have a valid, consistent way to measure performance.

The instructor should receive feedback from the learners on the training they receive. Formal written testing and performance evaluations are common methods of receiving feedback from recruits. Tests are used as an evaluation tool both to test skills (performance tests) and knowledge (written tests.) An example of a physical performance evaluation would be that the firefighter will don breathing apparatus in 45 seconds using procedures taught in recruit class (Figure 1.3). Performance tests are often conducted one on one, that is, the instructor evaluates only one student at a time. Because of the skill level required in the fire service, performance tests are vital and should be emphasized in training programs.

The fire service instructor should be continually evaluating and re-evaluating the training of the department. The fire service is not a static profession, and the instructor is responsible for

INSTRUCTOR'S INFLUENCE

- SELF-ESTEEM
- DESIRE
- DETERMINATION

Figure 1.2 Instructors have a great deal of influence on students.

keeping programs updated to conform to current practices. The instructor should look for ways that training may be improved. Observing the actions of firefighters at incidents can be helpful in determining if training is appropriate (Figure 1.4).

Figure 1.3 Performance tests are one way for the instructor to receive feedback. *Courtesy of Columbia Fire Department (Missouri).*

Figure 1.4 Skills learned in the classroom should be transferable to the fireground. *Courtesy of Boone County, Missouri Fire District.*

Because the fire service is constantly changing, training cannot be limited to just recruits. In addition to training firefighters in skills and knowledge needed in fire control, the instructor will establish courses for officers in more advanced and technical subjects, such as hazardous materials technician training. Seasoned firefighters can benefit from training by learning new methods and procedures. State and regional training programs, such as those offered by colleges and state training agencies, should not be overlooked as a valuable addition to local training.

Workshops or seminars in specialized training areas can benefit a department that may not be equipped to handle specialized training. Whether training recruits or seasoned firefighters, the instructor should not hesitate to seek the help of other qualified instructors, outside agencies, or specialists.

The fire service instructor must have many qualities: a thorough understanding of the subject matter, both from experience and knowledge; the ability to communicate effectively; and an understanding of the principles of teaching (Figure 1.5).

Figure 1.5 Skill and knowledge of fire fighting practices are essential, but a good instructor must also know how to teach.

CHARACTERISTICS OF GOOD INSTRUCTORS

The belief that good instructors are natural leaders and teachers may be true to an extent. Good teaching methods, however, can be practiced and learned. The attributes of a successful instructor cannot be scientifically analyzed, but certain techniques and behavior patterns influence successful teaching. Certain qualities contribute to the effectiveness of a good instructor (Figure 1.6).

An Ability to Understand and Work Well with People. An instructor must have a basic desire to understand the particular needs of the students. A student/teacher relationship built on mutual respect, confidence, and rapport is a positive aid to learning

Figure 1.6 The characteristics of instructors add to the effectiveness of their instruction.

(Figure 1.7). The instructor must also be able to work well with others in the department. The successful instructor should be fair and impartial toward all students.

Figure 1.7 An instructor must have a basic desire to understand the needs of the students. *Courtesy of Columbia Fire Department (Missouri).*

A Desire to Teach. Although instructors can improve their teaching ability through experience and study, their desire to teach will affect their efforts. As with any profession, desire and motivation contribute to the quality of the job performed.

Competence in Subject. An instructor must know the subject and related areas. A good instructor must have the ability and initiative to supplement knowledge. Keeping informed of problems, changes in procedures, and new concerns of the fire service requires that the instructor review the teaching material often and examine its relevance.

Enthusiasm. Enthusiasm is contagious. The instructor should follow through on student questions and project a high degree of interest in the subject matter. Citing personal experience and applications usually gives the students a clearer idea of the subject and aids retention.

Motivation. Motivation is the desire and determination to achieve goals. In order to be motivated, recruits must have a clear idea what is expected of them and how they are to perform. It is the job of the instructor to communicate these goals to the learner. Motivation includes giving people a purpose and a goal to achieve. The desire for recruits to succeed must be communicated effectively in order for it to translate into motivation. The enthusiasm and attitude of the instructor is vital in motivating students. Motivation, like enthusiasm, is contagious. The fire service is a team effort, not an individual effort, but individual members can do much to motivate each other.

Ingenuity and Creativity. A successful instructor understands that a technique suitable for one group may not be suitable for another. Approaches to different audiences must be different. Instructors demonstrate their creativity by developing or using training aids to supplement material, discovering other than traditional means of presenting material, improving material, and improving methods of measuring progress.

Empathy. Empathy is often said to be "putting yourself in another person's shoes." Empathy is the ability to understand the feelings and attitudes of another person. The instructor should be able to understand the students' point of view. An instructor should have a sincere desire to help the students learn and should never adopt a condescending or superior attitude.

TRAPS TO AVOID

Since an instructor is constantly observed and evaluated by the students, the impression students form of the instructor will affect their response and learning initiative. If students develop a negative attitude toward the instructor, learning will be inhibited. Listed below are factors that can produce negative impressions in students. The instructor should avoid the following pitfalls (Figure 1.8):

Bluffing. Bluffing an answer to a question will quickly destroy an instructor's credibility. No instructor can know all the answers to all possible questions, and students realize that fact. Students expect honesty and want an instructor who is willing to assist in finding answers to their questions. When a question is asked that cannot be answered, the instructor should not be embarrassed or shaken. Below are guidelines for effectively handling this type of situation.

- If the question is proper and pertinent to the class, promise to find the answer and do so.
- If the question is of interest only to the person asking the question, tell the student where to find the information.
- If there is no exact information or answer to a particular question, refer the student to related information.
- If the question refers to advanced material that will be covered in more detail later in the course, answer briefly and announce that it will be covered later. Make a note to refer to the question when the material is presented.

Figure 1.8 Instructors must be aware that they are constantly evaluated by their students.

Sarcasm. The instructor who uses sarcasm toward one student places the entire class on the defensive. The emotional reaction to sarcasm blocks effective communication and subsequent learning. Even if a student is sarcastic toward the instructor, the instructor must not retaliate in kind. Evaluate the situation and use a mature approach in dealing with it.

Complaining. Although few instructors work under ideal circumstances, airing complaints to students accomplishes nothing positive and can create negative impressions. Students have much more respect for an instructor who can present an interesting class with enthusiasm and optimism than they do for a chronic complainer. Students have little control over the teaching environment, and class time should not be taken up by apologizing or making excuses for a situation.

Comedy. Humor creates interest and releases tension in the classroom, but should be used only occasionally and always in good taste. Never tell a joke or story that has racial, religious, sexual, or ethnic connotations. Humor can be used to relax the class and to make it less formal and more interesting. Humor can be used to add emphasis to a subject if the instructor keeps teaching, not entertaining, in mind.

Bullying. The instructor who views the trainer's role as a pusher, rather than as a teacher, will not stimulate students to reach their potential. Bullying is usually an expression of fear or frustration and causes people to react fearfully and contemptuously.

COMMUNICATION

The ability to communicate effectively is an important quality of a good instructor. The fire service instructor must not only have knowledge and ability but also must be able to communicate that knowledge and ability to students. Communication is much more than just "talking," it is a process that explains the interaction of individuals. Being an effective communicator takes practice; instructors must not assume that they are good at talking with or in front of people and leave it at that.

Everyone can benefit from understanding the process of communication and evaluating their own communication skills. The vital point in communication is that it must be understood. An instructor may have no problem lecturing to students, but it is essential that those students clearly understand the information and be willing to ask questions and involve themselves in the communication process.

The communication process exists in many forms, but all communication accomplishes the task of exchanging information. Regardless of the form used, the communication process consists of five essential elements:

- A person to send the information — the "sender"
- The actual message
- A method for the information to be transmitted
- A person to receive the information — the "receiver"
- Feedback from the receiver to the sender that the information was received and understood

The Sender And Receiver

The communication process begins when the sender perceives that another person must perform in a given manner to help meet the needs of the sender (Figure 1.9). The sender must convey to the second person (receiver) the intended message of the sender. The sender must develop a message and method of transmission that the receiver can understand.

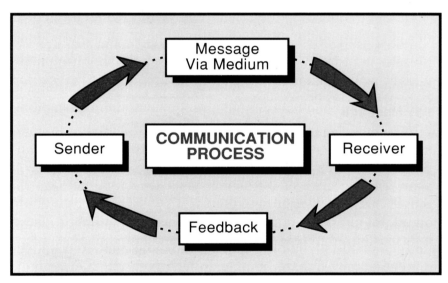

Figure 1.9 Communication is accomplished when the receiver confirms the intended message has been received.

Communication is an active process that involves much more than just a message and its sender. Communication is aimed at creating understanding between the sender and the receiver. Each person in a conversation acts as both a sender and receiver (Figure 1.10 on next page). Feedback from the receiver to the sender gives the sender an indication of how the receiver is perceiving the message. An important fact to keep in mind when communicating is that the sender and receiver may perceive the situation being discussed differently because of their differing backgrounds. We all use our background and past experiences to interpret what we see and hear. Because of this, two people rarely interpret a statement they have heard in exactly the same way. Difference in perception of a situation is often a barrier to communication. This means that the sender must consider the feelings and personal background of the receiver. The sender

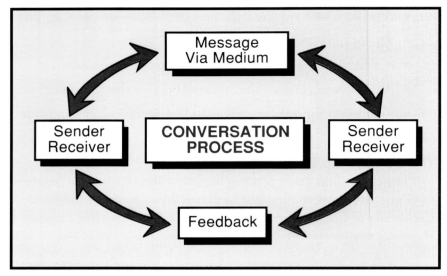

Figure 1.10 The communication process is a conversation when each person becomes both the sender and receiver.

must develop thoughts into an understandable package. This package is called the message. The message is the thought that is transmitted. The method of transmission is called the medium.

The Message

The message communicated between the sender and receiver must convey the intended meaning and be in a form that is understood. This process is sometimes harder than it may appear. The objective of the sender is to develop a message that conveys what is wanted or expected. It serves no purpose to send or request information that is neither needed nor wanted. Useless information only clutters or confuses the communication process.

The receiver must be able to clearly understand what is expected or is being asked. If this essential task is not accomplished, the entire communication process begins to falter.

The Medium

The medium of communication is the channel by which the message is delivered between the sender and receiver. The selection of a communication channel will depend upon many factors such as available time, purpose, language, or the information being delivered. For example, it may be acceptable to verbally communicate a message on the fireground rather than trying to send a written message. One method is appropriate for the situation; the other may deliver the message, but not in a timely fashion. Other examples of communication channels, in addition to spoken words or written messages, are physical actions of an individual, visual displays for training, or the use of various tones and sounds. Each may appropriately convey the message in the most effective way possible. The end result should be an understanding of the message by the receiver.

The instructor must always consider the possibility of a language barrier or misinterpretation of terminology. The best way to keep this from happening is to use simple, direct words in brief, uncomplicated messages. This is particularly important during emergency situations when excitement and activity levels are high. The channel of communication and the design of the message are key considerations that the instructor must learn to address when communicating with others.

The scope of medium in the communication process includes the nonverbal communication commonly called "body language." When we communicate orally, the receiver often "hears" what we don't say louder than our words. Our gestures, facial expressions, stance, and even how we dress sometimes speak louder than our voice. What is communicated nonverbally may not be what we had in mind, but it is usually a large part of how the message is interpreted.

Communication is not always verbal and the instructor will use many different mediums to convey messages. An important medium used for communicating information is through written words, graphic displays, or the use of commonly accepted symbols. In this form of communication, the receiver must visually identify the message, interpret the meaning, and respond or provide feedback that the message was understood. Examples of written forms of communication are:

- Letters or other correspondence
- All instructional materials
- Symbols used in pre-incident planning
- Training aids (graphics, charts, and photos)

Each example of written communication must convey the information that the sender wants the receiver to understand and must be in a form that is appropriate for the message.

Listening

Communication is an active process, and the individual who is not speaking must remain actively involved by listening. Listening is just as important an aspect of communication as speaking is. The purpose of listening is not to learn how to get others to listen to us, but how to listen *ourselves*. Listening to others takes up a good part of our time every day. Research has shown, however, that the average person takes in only a fraction of what is heard. Fortunately, good listening habits can be learned.

Some people merely *hear*; they do not listen. There is a very good chance these people will not understand the meaning of the message being sent. Understanding a message requires that the listener take an active role in the communication process. Alert facial expressions and posture are indications of a good listener. Questions and comments from the listener show interest and encourage the speaker to expand on what is being said. These actions give the speaker feedback on how the message is being received.

When listening, an individual must do more than just hear the speaker's words: the listener must understand those words. Words have different meanings for different people, so the listener must attempt to understand the speaker's intent. Listening with understanding involves interpreting word meanings, that is, listening with understanding involves empathy. Empathy is the ability to put oneself in another's place, and it is an essential skill for the fire service instructor.

The instructor should attempt to communicate with all students on the same level. Bias or prejudice should not be a barrier to communication. The instructor and other members of the class must not discriminate against someone because of age, sex, race, or religion. Every student in the class has the right to be heard.

A good environment for listening is important, especially in a classroom situation. Absence of loud background noise, a comfortable temperature, and the expectation of both sender and receiver that the message will be important enough to merit attention make for a good listening environment. The instructor must constantly be aware of the listening environment or run the risk of both not being heard and not hearing others.

Speaking Techniques

A good communicator is an effective speaker. An instructor's voice should be clear, distinct, and not monotonous. Voice fluctuations add variety and can be used to add emphasis. Speaking speed should be slow when beginning a new class and can gradually be increased as the students become familiar with the instructor's voice. Slow down on important items and speak especially slowly when students are taking notes. If items are particularly important, the instructor may want to restate the point to add emphasis.

When speaking, the instructor should avoid using improper grammar and slang words that students may not understand. The instructor should also avoid using uncommon or highly technical terms without explaining their meaning. Words that are common to the instructor may not be common to the students. The instructor should pause when speaking to allow students a chance to comprehend what is being said. Speaking in front of people can cause stress and anxiety. The instructor should try to relax when speaking and keep a conversational tone.

Personal mannerisms are usually distracting and should be avoided. Some examples of undesirable mannerisms include:

- Pencil, toothpick, or match chewing
- Frowning or glowering
- Foot tapping
- Floor pacing
- Finger snapping
- Profane language
- Playing with chalk
- Repetition of words
- Cleaning or biting fingernails
- Pulling or adjusting clothing
- Jingling coins or keys
- Clock watching
- Playing with jewelry
- Excessive use of "I," "OK," "you know," and so on
- Looking away from the students (for example, at floors or walls)

One of the most desirable traits an instructor can develop is that of looking directly at the students. Eye contact reinforces the feeling that the instructor is interested in the students and concerned that they understand the material. Gesturing can also be effective if properly used to draw mental pictures or to emphasize key points (Figure 1.11).

Figure 1.11 Gesturing can be an effective way of emphasizing an important point.

INSTRUCTOR DEVELOPMENT

A competent instructor is never fully trained and should be willing to receive, as well as to give, instruction. The effective instructor will constantly strive to improve teaching approaches and techniques. The fire service is constantly changing and the instructor must stay aware of new improvements or developments. Workshops, seminars, and self-study can be valuable aids to the instructor in teaching. Instructors should keep updated on fire service practices not only for their own benefit but for the benefit of recruits and others in the department (Figure 1.12).

Figure 1.12 Instructors should keep updated on what is happening in the fire service.

IMPORTANCE OF INSTRUCTION

The fire service is different from many professions in that the consequences of wrong actions can be devastating. Firefighters face incidents that are potentially life-threatening on a continual basis and must be able to react to those incidents effectively. The fireground is not the place to second-guess teaching techniques and methods. The safety of the firefighters and others depends on the quality of training the firefighters receive.

The satisfaction that instructors receive from doing a good job cannot be measured. The knowledge that their dedication and devotion to the fire service may have helped save the lives of victims, as well as firefighters, is a great reward. When recruits graduate or an apparatus operator is promoted to officer rank, an instructor shares that promotion because of the assistance provided to the promoted individual through the training program.

The effective instructor should be internally motivated to do a good job. Although recognition from the department is important, the personal satisfaction from teaching is vital for the instructor (Figure 1.13). Without motivation and satisfaction, it is unlikely that an instructor will be effective.

SUMMARY

The fire service instructor plays a vital role in the fire department. The instructor is responsible for planning and conducting training as well as ensuring that performance standards are met. Effective instructors will have characteristics that enable them to be successful at teaching. These characteristics include the ability to understand and work well with people, a desire to teach, competence in the subject, enthusiasm, motivation, ingenuity and creativity, and empathy.

Effective instructors will be honest with the students and be able to communicate effectively. The instructor should avoid bluffing, sarcasm, complaining, and comedy while teaching. These only serve to undermine the credibility of the instructor. The ability to communicate effectively is essential to teaching. Communication is a process that involves much more than just talking, and fire service instructors should understand it thoroughly. The elements of the process of communication are sender, message, channel, receiver, and feedback. Listening is an important aspect of communication and instructors must be able to effectively listen to students. An effective communicator will also be an effective speaker and instructors should be aware of their speaking style. Instructors must be knowledgeable about the fire service and be able to relay that knowledge to students. A good instructor is never fully trained and must constantly strive to improve teaching approaches and techniques.

Figure 1.13 Instructors should take pride and achieve personal satisfaction from their job.

SUPPLEMENTAL READINGS

Bachtler, Joseph. *Fire Instructor's Training Guide.* 2nd ed. New York: Fire Engineering, 1989.

Granito, Anthony R. *Fire Instructor's Training Guide.* New York: Dun Donnelly Publishing Corporation, 1972.

Knowles, Malcolm. *The Modern Practice of Adult Education: Andragogy Versus Pedagogy.* New York: Association Press, 1970.

Laird, Dugan. *Approaches to Training and Development.* 2nd ed. Special material by Peter R. Schleger. Reading, Mass.: Addison-Wesley Publishing Company, Inc., 1985.

Miller, W.R., and Homer C. Rose. *Instructors and Their Jobs.* Chicago: American Technical Society, 1975.

Veri, Clive C., and T.A. Vonder Haar. *Training the Trainer.* St. Louis: Extension Division, University of Missouri-St. Louis, 1970.

Safety and the
Fire Service Instructor

This chapter provides information that addresses performance objectives in NFPA 1401, *Fire Service Instructor Professional Qualifications* (1987), particularly those referenced in the following sections:

NFPA 1401

Fire Service Instructor

3-4.1 (b)

3-7.2 (b)(c)

Chapter 2
Safety and the Fire Service Instructor

Safety is a paramount subject in fire service instruction and one of the most needed, considering the severity of the problem. For many years instructors have been teaching a wide variety of subjects, but safety has not been stressed as strongly as it should be in many departments. This deficiency is quite evident when looking at accident statistics. The fire service has the dubious honor of having one of the highest accident rates in any employment field.

The fire service is an inherently dangerous profession, but the risks involved in fire fighting can be reduced if departments take a proactive role in safety. Departments should set up safety programs and stress safety throughout the department. Everyone in the department, from beginning recruits to seasoned firefighters and administrators, should be involved in the safety program. Safety should be emphasized in training and the instructor should devote enough time to adequately cover the subject of safety when discussing lessons. Seasoned firefighters should not take their experience for granted but should continue to focus on safety and accident prevention.

The fire service instructor can have a sizable impact on safety. The instructor is the one person with which new recruits always interact, and he or she sets the stage for the remainder of the recruit's career. The fire service instructor must have a significant positive effect on safety in the organization (Figure 2.1 on next page).

An effective safety program includes appointing a safety officer, maintaining accurate accident and injury records, and having additional training or ongoing training programs.

Safety should begin in the fire station, where firefighters spend a great deal of time. The fire station should be checked for

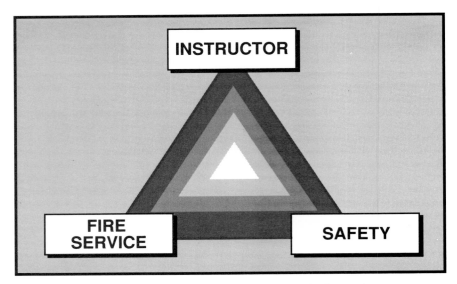

Figure 2.1 Instructors can have a positive impact on safety in the fire service.

conditions that may lead to accidents. Remember — accidents do not happen only on the fireground. Areas that are possible problems include slippery surfaces, stairs, slide poles and slides, electrical outlets, and office furniture and equipment.

Fire fighting involves a certain degree of risk, and safety at incidents is very important. Firefighters should be trained and qualified for the type of incident they are involved in controlling. Incident safety includes wearing proper protective clothing and self-contained breathing apparatus, as well as the proper use of other specialized protective equipment.

Firefighters depend on their vehicles and equipment to effectively fight fires. Because of this, they must be maintained in good condition and handled with care. Vehicles or equipment that are not in good condition are hazards to those who use them.

Safety cannot be overemphasized and departments can work to reduce the number of accidents and injuries. Implementing a safety program and actively involving members in the program is a good start. The instructor should realize the importance of safety and give it adequate coverage in classes. Safety procedures and precautions are not a luxury to firefighters — they are a necessity.

The instructor must stress safety from the beginning and make sure it is the highest priority in everything the student does (Figure 2.2). Students will follow what they are shown, so be sure that everything is done in the safest manner possible. Students can figure out the wrong ways to do something all by themselves; the instructor's job is to teach them the correct, safe way.

The first item to be taught should be from the written department safety policy. This policy should be presented by the chief, or management, to show the support they give it and how much

Figure 2.2 Safety should always be a primary concern in the fire service.

it means to the department. This will also show that the instructor has the backing of management.

The policy should state that accidents are going to be investigated, not to cast blame, but to determine what went wrong. Accident investigations can be an excellent tool for the instructor to use in revising what is being taught.

ACCIDENTS

Accidents usually result from unsafe acts or physical situations. An accident can be defined as an unplanned, uncontrolled event that results from unsafe acts of persons and/or unsafe occupational conditions, either of which can result in injury. Typically, accidents occur because an established safety procedure is not followed or not established. Safety is not just a set of procedures; it is an attitude that must be continually stressed until the habits of safety are fixed in the learner (Figure 2.3). Safety should be further reinforced during in-service programs.

Regardless of how difficult it may be to determine or detect, the cause of an accident is something definite. When an accident occurs, it is an indication that someone has failed to perform in a safe manner or that an unsafe condition existed. Accident cause and safety procedures must be evaluated after any training program.

Figure 2.3 Safety procedures must become a habit to firefighters if accidents are to be prevented.

Human Factors

It is sometimes difficult to understand why people create or tolerate unsafe mechanical or physical conditions or why they

HUMAN FACTORS

- Improper Attitude
- Lack of Knowledge or Skill
- Physically Unsuited

Figure 2.4 Accidents can be caused by human factors.

commit unsafe acts. It has been found, however, that people do act in an unsafe manner due to several factors that may be referred to as "Human Factors" (Figure 2.4).

In industrial accident analyses, it has been found that accidents do not distribute themselves by chance; accidents happen frequently to some people and infrequently to others. Those individuals who, because of mental, psychological, or physical defects, fail to control a hazardous situation will become involved in accidents more frequently.

Three factors that can contribute to accidents and are attributable to human defects are discussed below:

Improper attitude — Includes willful disregard, recklessness, irresponsibility, laziness, disloyalty, uncooperativeness, fearfulness, oversensitivity, egotism, jealousy, impatience, obsession, phobia, absentmindedness, excitability, inconsideration, intolerance, or mentally unsuited in general. Readjusting faulty attitudes or personalities can lead to accident reduction.

Lack of knowledge or skill — Includes insufficiently informed, misunderstandings, not convinced of need, indecision, inexperience, poor training, and failure to recognize potential hazards. These defects can be corrected by properly training or retraining the firefighters.

Physically unsuited — Includes hearing, sight, weight, height, illness, allergies, slow reactions, crippled, intoxicated, or physically handicapped in general. In correcting these physical disabilities, accident rates can often be reduced.

Accident Investigation

Accidents do not just happen; they occur through a logical and predictable sequence of events. When accidents occur, an investigation should follow to determine exactly what happened (Figure 2.5). Investigations must be started toward "fact" finding and

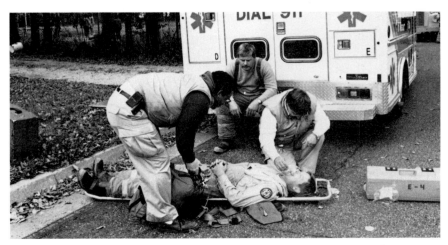

Figure 2.5 Accidents must be investigated to determine exactly what happened.

not "fault" finding. They should be done with objective determination, free of personal feelings toward any of the principals in or out of the fire department. Several purposes of accident investigation include:

- Avoiding loss of human resources and equipment

- Ensuring better cost effectiveness in the use of manpower and equipment

- Improving the morale of department personnel and the community

- Determining the change or deviation that caused the accident

- Determining the hazardous conditions to which firefighters may be exposed

- Directing officers' attention to accident causes

- Determining facts that could have a legal impact on an accident case

Fire service instructors should work with the safety officer in conducting investigations of accidents occurring within their supervision since they will usually be on the scene and have more information concerning the details of the accident.

Accident Analysis

After all data has been obtained and compiled, it must be broken down and analyzed. Reviewing existing records may reveal a significant area of concern. Many departments easily become complacent in their assessment of their activities. Careful record keeping often reveals minor situations that indicate the need for corrective action. An inconvenient scratch or bruise may not be disabling, but can indicate a need for corrective action. To the reviewing officer, a number of small scratches or blisters may indicate a significant trend that can be corrected before serious injuries occur. Analysis of circumstances pertaining to accidents can:

- Identify and locate principal sources of accidents by determining the materials, machines, or tools most frequently involved with accidents and job-producing injuries.

- Disclose the nature and size of the accident problem in various companies.

- Indicate the need for engineering revisions by identifying the unsafe conditions of the various types of equipment.

- Identify problems in operating procedures and processes contributing to accidents.

- Disclose unsafe practices that might necessitate additional training.

- Identify improper personnel placement where inabilities or physical handicaps contribute to accidents.

- Enable company officers to better use their time for safety work by providing information about hazards and unsafe practices in their company.

- Permit an objective evaluation of the safety program's progress.

Accident analysis can reveal the need for additional training in specific procedures. Individuals may act with good intentions yet not understand the consequences of their actions. This situation can result in accidents and injuries. Individuals who perform unsafe acts are usually not aware of the safety factor until an accident occurs.

It is up to the training instructor to make sure that all training is done in the safest manner possible for everyone involved (Figure 2.6). All training exercises should be looked at with safety in mind from the beginning to the end, whether on the fireground or in the training center.

Figure 2.6 The training instructor is responsible for ensuring safety on the training ground.

There are several government agencies that are responsible for helping to ensure a safe workplace. The two primary agencies concerned with safety in the workplace are the National Institute for Occupational Safety and Health (NIOSH) and the Occupational Safety and Health Administration (OSHA). Both agencies were created by the Occupational Safety and Health Act of 1970. The responsibilities of NIOSH include investigating, researching, and evaluating safety and health hazards in the workplace. The responsibilities of OSHA include setting and enforcing workplace safety and health standards. OSHA has the authority to issue citations and fines to enforce safety and health standards.

There are several NFPA standards that relate to safety and the performing of live evolutions with which the instructor should be familiar. NFPA 1402, *Standard on Building Fire Service Training Centers (1985)* lists guidelines that should be followed when building training facilities, including burn buildings, smoke buildings, combination buildings, and outside drillground activities.

NFPA 1403, *Standard on Live Fire Training Evolutions in Structures (1986)* lists guidelines that should be followed during live training evolutions. It includes information on structures, fuel materials, safety, instructors, and reports and records.

NFPA 1500, *Standard on Fire Department Occupational Safety and Health Program (1987)* contains information on the guidelines that fire departments should follow in order to ensure the health and safety of firefighters and help prevent accidents and health problems.

STRUCTURAL FIRE TRAINING

Structural fire fighting buildings at fire department training facilities provide an opportunity to conduct a large number of interior training fires in a limited space and in a limited time. Fire frequency is governed by the type of fuel burned. This, in turn, dictates the cleanup, refueling, and re-use cycle. The convenience of being able to produce training fires quickly and in relative safety is offset to some degree by a lack of authenticity. Training ground structural building fires may be "like real" but they cannot, due to lack of structural involvement, be real (Figure 2.7).

Live fire training conducted in suitable buildings available for demolition provides the realism missing in training center fires.

Figure 2.7 Training ground structural building fires can aid training and are relatively safe and easy to use.

The sights, sounds, and sensations the trainee experiences are real. While this level of realism provides excellent training, it obviously carries with it most of the hazards of interior fire fighting at an actual emergency. Training sessions must, therefore, be planned with great care and supervised closely by qualified instructional personnel.

Building construction, building condition, exposures, terrain, water supply, and a multitude of other variables affect the safety of participants. The following is a list of some of the details that must be considered. It is not intended to be exhaustive, nor can any list cover every contingency. The on-site judgment of qualified personnel must be the final factor in determining the safety of any specific situation.

Site Inspection

- Building construction, detailed list
- Building layout, detailed sketch
- Building condition, detailed examination
- Exposures, type and distance
- Prevailing wind
- Downwind hazards, type and distance
- Wires, type and location
- Trees and other obstructions
- Vehicle access and spotting areas
- Stability of access roads and parking areas
- Water supply, type and accessibility
- Anticipated spectator traffic
- Outbuildings, trees or other items to be protected and saved

Permits, Documents, and Notifications

- From owner:
  - Certification of ownership
  - Permission to burn
  - Certification no insurance is in force
  - Certification no liens or encumbrances apply
- Environmental health agency (state and/or local)
- Forestry Department (wildland areas)
- Razing permit (where required)
- Notification to water department (if necessary)
- Property liability insurance obtained covering damage to other property

- Notification to central fire dispatch
- Notification to affected police agencies
- Authority to block roads (if required)
- Assistance in traffic control requested (if required)

Building Preparation

- Building correctly identified and marked
- Floors made safe
- Window openings closed
- Necessary doors in place
- Loose wallpaper removed
- Linoleum removed
- Stairways made safe — railings in place
- All chimneys undercut
- All holes in walls and ceilings patched and sealed
- Openings made in end gables for observation and fire extinguishment
- At least one 4 x 4-foot (1.2 m by 1.2 m) hole made in each separate roof area for ventilation
- All inside debris cleaned up and rooms swept
- All outside debris cleared away from entrances and areas of egress
- Porches and outside steps made safe
- Outside grounds prepared for vehicles
- Cistern, cesspool, well, or other ground openings protected or filled
- Fuel tanks checked
- Utilities (electric, phone, gas, water) disconnected
- Insect hives and toxic weeds removed
- Building secured

Training Preparation

- Goals and sequence of training established
- Water supply planned, apparatus assigned
- Hose layouts planned (primary and backup hoselines from different pumpers)
- Fuel supplies/burn materials obtained and placed
- Use of gasoline and other low flash point fuels prohibited
- Qualified officer-in-charge assigned

- Qualified instructors assigned
- Qualified safety officer assigned and authorized to take any necessary action
- Downwind ember patrol planned and assigned
- All personnel fully briefed

Training Operations

- Exposure protection in place and operating
- All participating personnel in full gear with breathing apparatus, gloves on, boots up, and ear flaps down
- Total entry and exit head count maintained at all times
- Fires ignited only on authority of officer in charge
- Safety officer in appropriate position to observe conditions
- Strict discipline enforced by all instructional personnel
- All safety and discipline measures maintained until termination of session

Before Training Evaluation

- Physical makeup and fire travel potential
- False alteration problems
- Possibilities of structural collapse
- How will heat vents affect smoke travel
- Have stairways and safety features been removed
- Has entire structure been searched for flammable liquids or explosives
- Fire loading vs. fire building
- Members of a department covered by insurance and Worker's Compensation
- Able to use SCBA
- Able to use ladders
- Able to use forcible entry tools

Incident Scene

- Potential personnel hazards
- Proper protective clothing and how to dress
- Building condition and type of construction
- Rescue problems that may occur during exercise
- Insufficient personnel for the task
- Hazardous materials

- Animal and insect problems — raccoons, rats, cockroaches, ants, snakes
- Weather extremes (hot and cold)
- Irate neighbors

Training Problems

- Refueling for inside and outside training
- Portable training aids
- Plans in case things go wrong
- On-scene safety officers who care about the welfare of personnel and are allowed and expected to do their job
- Fires started when training officer determines everything is ready and not before
- Complete discipline until exercise is finished

Burn Day

- Student registration
- Establish a command post
- Brief students and instructors
- Weather report and changes that may occur
- Problems that may occur during breakdown or lunch

Station

- Surface problems, running in stations
- Vehicle maintenance
- Equipment maintenance
- Office safety

Summary

The fire service can reduce the accident rates in fire fighting by taking an active role in safety. The fact that fire fighting involves risk that cannot be avoided should not be a justification for high accident rates. Some accidents can be avoided if correct safety procedures are followed. Safety should be a primary part of every lesson. With just some common sense additions to the current curriculum and a realization that the best teaching is by example, today's fire service instructor can have a real impact on tomorrow's accident statistics in the fire service.

SUPPLEMENTAL READINGS

Firefighter Occupational Safety. Stillwater, Okla.: Fire Protection Publications, Oklahoma State University, 1979.

Soros, Charles. *Safety in the Fire Service.* Boston: National Fire Protection Association, 1979.

NFPA 1500 Standard on Fire Department Occupational Safety and Health Program. Quincy: National Fire Protection Association, 1987.

NFPA 1402 Guide to Building Fire Service Training Centers. Quincy: National Fire Protection Association, 1985.

NFPA 1403 *Standard on Live Fire Training Evolutions in Structures.* Quincy: National Fire Protection Association, 1986.

NFPA 1501 Standard for Fire Department Safety Officer. Quincy: National Fire Protection Association, 1987.

NFPA 1410 Training Standard on Initial Fire Attack. Quincy: National Fire Protection Association, 1988.

Legal
Considerations

This chapter provides information that addresses performance objectives in NFPA 1401, *Fire Service Instructor Professional Qualifications* (1987), particularly those referenced in the following sections:

NFPA 1401

Fire Service Instructor

3-2.3

5-4.5

Chapter 3
Legal Considerations

Fire service training officers and instructors are important members of the human resource management team. In this capacity they assume significant responsibility for the career opportunities of many individuals (Figure 3.1). Fire service training officers and instructors are in a professional position of high trust that has critical legal implications if they fail to perform adequately.

Society is becoming increasingly complex and dependent on the law. As a result, instructors are finding that the results of their actions in carrying out their routine duties are having a significant legal impact on both themselves and their organizations. Hence, instructors must have a basic knowledge of the law.

Figure 3.1 Instructors have a responsibility toward the fire service and the individuals with which they work.

This chapter provides an overview of selected legal considerations that need to be made when hiring personnel, promoting personnel, and developing curriculum. It contains sections on liability, basic definitions and concepts, equal employment opportunity and affirmative action, copyright, and right of privacy.

DEFINITIONS AND CONCEPTS

Law

The fire service instructor is influenced every day by a number of laws, ordinances, and regulations (Figure 3.2). One common misconception regarding the law is that it is written down in a neat set of statutes in one large volume. Nothing could be further from the truth. The "law" comes from many different sources and a great deal of it is inferred from interpretive decisions made by judges and administrative hearing officers.

The great bulk of American Law has its roots in the English Common Law. The adaptation from that time to the present day carried some positive and some negative legal connotations. The doctrine of sovereign immunity stems from the personal position of the English king, and basically meant that the king, and hence, the government, were immune from legal suits.

The Common Law doctrine carried over to American Law and had the effect of holding any federal, state, or local governmental body immune from liability in tort. Tort is a wrongful act for which a civil action will lie. The practical effect was that any governmental body or agency, or employee thereof, was immune from liability for any action taken, negligent or otherwise, in an official capacity.

In 1946, the United States Government waived its immunity from liability in tort and provided for the litigation of tort claims in the federal courts. Until recently, most states were immune from suits for tortious injury to persons or property. In the past few decades, however, the doctrine of sovereign immunity has undergone considerable erosion, legislative modification, and, in some cases, outright abolition.

Today, the continuing trend away from sovereign immunity is clear, and tort and/or negligence liability clearly exists in all but a few jurisdictions. An important fact to note is that the individual employee as well as the governmental agency may now be held liable in tort.

Tort Liability

A tort is a civil wrong or injury. The main purpose of a tort action is to seek payment for damages to property and injuries to individuals. Liability is a broad, comprehensive term describing a person or organization's responsibility in the law. This responsibility implies that if a wrong has occurred, a person or an

INSTRUCTORS ARE INFLUENCED BY

- LAWS
- ORDINANCES
- REGULATIONS

Figure 3.2 Instructors should be aware of the legal aspects of the fire service.

organization must respond to legal allegations. By definition, the best way to minimize the possibility of litigation is to avoid negligence.

The fire service instructor is faced with a variety of situations that could result in liability. Instructors should be aware of these situations and take steps to prevent liability from occurring (Figure 3.3). Liability can result if:

- The instructor is negligent in training students, including exposing students to unnecessary risks or failing to warn students of the potential danger of an exercise.

- A student receives an injury, after training, but the injury was caused by poor or incorrect instruction. This is difficult to establish; however, it can be established if needed safety warnings were not included as part of the formal instruction.

- An injury to a third party was caused by poor or incorrect instruction.

- There is a misrepresentation of qualifications or benefits of training, including making claims for training that cannot be supported. An example would be instructors teaching in areas in which they have no qualifications.

- There is improper supervision of interns or trainees. There must be specific guidelines governing the quality of the trainee's work.

- There is improper communication including libel, slander, or breach of confidentiality.

LIABILITY
- Negligence in Training
- Incorrect Instruction
- Improper Supervision
- Improper Communication

Figure 3.3 Exposure to liability can be reduced if the instructor ponders the need for precautions.

CONDITIONS FOR LIABILITY

There are several conditions that must be present in order for liability to occur. Each of these conditions must be met to show liability. The plaintiff must show:

- The defendant owed a legal duty which was the minimum required to protect the learners from unreasonable risk.
- The instructor failed to act according to that duty.
- There is a causal relationship between the action or inaction of the instructor and the resulting injury.
- The plaintiff sustained damages.

The first element, the matter of duty, is relatively easy to establish in a fire department related tort. The fire service instructor owes a legal duty to the students to protect them from unreasonable risk during training (Figure 3.4). The key word here is unreasonable. Fire fighting is dangerous and involves a "peculiar risk" that cannot be avoided. If the risk the instructor asked of the students was dangerous beyond that peculiar risk or the students' level of ability, then the risk would be considered unreasonable.

Likewise, the fact that the plaintiff suffered damages is readily established in accident cases. For example, damages may take the form of property repairs or replacement, medical expenses, or lost income. The dollar value of damages suffered, however, is an issue that may involve a considerable portion of a court proceeding.

CAUSATION

The question of causation is more difficult to establish. A proximate cause is one that in a naturally continuous sequence produces the injury, and without which, the result would not have

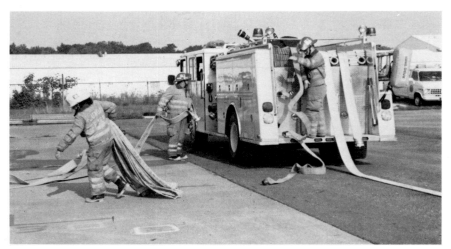

Figure 3.4 The instructor should take steps to prevent the students from unreasonable risk during training.

occurred. Proximate cause can be a sticky legal question. A showing of responsibility for creating the original action can be enough in many cases to establish proximate cause.

Most accidents are the result of multiple factors. The proximate cause issue may be downplayed by a jury where the injuries are substantial or emotions run high, such as when a child is badly injured. When a jury is searching for a "deep pocket," (the litigants with the most money) they may be satisfied with a minimal linkage, negating the proximate cause criterion. For example, in some instances, it may be sufficient to show that the procedure employed was not in accordance with acceptable agency standards.

NEGLIGENCE

Breach of legal duty, the second element of a tort, is the major issue in most tort liability cases. Negligence may be defined as the failure to act in a "reasonable and prudent" manner, such as failing to show appropriate diligence and care. If the person possesses a greater amount of expertise, the duty is proportionately greater. For instance, the standard of care for a fire service instructor would be that which reasonable, prudent, and careful fire service instructors would be expected to possess (Figure 3.5). The essence of negligence is the adequacy of performance. There are two ways in which one can be judged negligent; wrongful performance (misfeasance), or not performing when some act should have been taken (nonfeasance).

STANDARD OF CARE

The critical issue in liability is the care with which the instructor's responsibilities are exercised. If conduct falls below a reasonable standard of care, the responsible persons and/or organizations may be held liable for injuries and damages that resulted from such conduct.

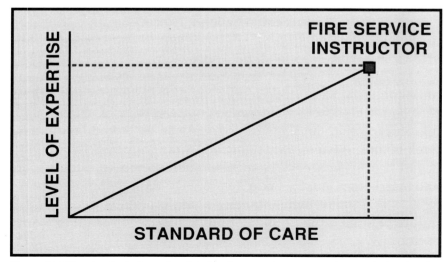

Figure 3.5 The standard of care is related to the expertise a person in a particular situation would be expected to possess.

There are factors that may limit one's ability to act. One has a responsibility to act in a reasonable manner, based on the information at hand and the resources available. When a potentially hazardous condition exists, the reasonableness of action must take into account the following factors, particularly when resources are not available to correct all such conditions:

- Gravity of harm posed by the condition

- Likelihood of harm

- Availability of a method and/or equipment to correct the situation

- Usefulness of the condition for other purposes

- Burden of removing the condition

Many items of information may be brought into court to aid in establishing the prevailing standard of care. One of the strongest types of evidence is the agency's own guidelines and policies. Regulations adopted by the agency may define in detail the minimum requirements. A reasonable person would follow such rules and orders.

Foreseeability

Since the concept of "reasonable and prudent" is situational, instructors must make every effort to foresee potentially dangerous or damaging situations.

The concept of foreseeability means that instruction should be based not only on dangerous conditions that may exist in training, but also should anticipate conditions firefighters might face on the job. As an example, fire control training in a burning structure would require such competencies as a knowledge of fire behavior, the use of self-contained breathing apparatus, and the ability to advance a hoseline. Most drill tower or academy instructors recognize these types of competencies. However, recent litigation involving public safety personnel suggests that training must be able to be generalized to all field situations. Thus, as technology and health issues have changed, the duties and tasks of firefighters have changed. Some examples of changes in technology include the complicated field of hazardous materials, lightweight building construction, and advances in rescue tools and equipment. Each of these developments, and many others, impact the day-to-day work of firefighters. Consequently, fire service training officers and instructors must prepare to properly train their personnel to handle these developments. Accidents occurring because of a failure to foresee the need for training may result in an exposure to liability.

Training officers and instructors should regularly conduct an occupational or job analysis that reflects the contemporary duties and tasks of the field. After extensive review of litigation involving vocational educators, it has been determined that courts have expected reasonable and prudent instructors to:

- Have a plan, namely a formally developed strategy to prevent injuries.
- Follow the plan.
- Provide for health and safety.
- Give proper instruction.

There are precautions the instructor can take to minimize the chance of becoming involved in a liability case. Some actions the instructor should take are listed below.

- Instruct and test all students in the safe operation of equipment.
- Maintain written objectives and document each training session.
- Check equipment for safe operating conditions regularly.
- Never leave the class unattended while students are engaged in a potentially dangerous area of instruction.
- Make sure trainees are physically fit and prepared for tasks.
- Provide students with a written description of the course and give an accurate representation of instructor qualifications.
- Respect student privacy.

LIABILITY

Personal Liability

Instructors owe a duty to both their students and the public. The duty owed to the public for reasonably safe care in all situations extends to all parties responsible for abating hazardous situations and delivering emergency care. This includes individual employees of public agencies and private contractors. Basically, all individuals have the obligation to conduct themselves in a manner that does not negligently cause harm or further harm to another person. An individual who violates this general duty of care can be sued for damages.

Instructors also owe a duty to the students they are teaching. They are responsible for ensuring that students have a reasonably safe learning environment. Instructors should take the necessary precautions, as discussed earlier, to limit the possibility of liability actions (Figure 3.6 on next page).

If a court or jury decides an individual is liable, then a judgment of damages can be returned against the individual. Recovery of punitive or exemplary damages may be one reason for suing an individual employee, especially where the public agency is immune from paying such damages. From a practical standpoint, however, employees are not often held responsible for payment of awards, particularly government employees. Because the individual's assets are so small as compared to those of the government, the "deeper pockets" will most likely be targeted for recovery of damages. Nevertheless, being named as a defendant in a lawsuit is a serious experience.

Figure 3.6 Instructors should make sure that students understand the correct procedures. *Courtesy of Columbia Fire Department (Missouri).*

Fire Department Liability

Fire departments must realize they are vulnerable to liability suits and take a proactive stance in dealing with liability. Fire service personnel work in an environment that is constantly changing; they must deal with different situations on a daily basis. The legal issues concerning fire service personnel are also changing. Legal considerations vary from state to state and departments must be aware of the laws surrounding their particular situation. These factors make it essential that fire department personnel be aware of possible litigation and take precautions to prevent them.

There are numerous situations where fire departments could be involved in liability cases. Vehicles, personnel, property, and incident scenes all have potential liability issues. The fire department could be sued by anyone who believes they have been wronged. Even if the department is cleared of wrong doing, it will incur the cost of defending the lawsuit.

There are precautions the department should take in order to minimize the chances of becoming involved in a lawsuit. The department must have a risk management program, which means it must make an effort to both understand and control risk to protect itself from losses. The department must conduct itself with a high degree of professionalism. Professionalism entails competence. To ensure personnel are highly competent, the department must adhere to performance standards. Having unqualified personnel working an incident is both dangerous and increases the chances of being involved in a liability case.

Stressing safety is an important element in minimizing possible liability cases. Safety should be emphasized at all times—during training, in the station, and at the incident scene. All equipment and apparatus should be properly maintained and inspected on a regular basis. For more information on safety, refer to the IFSTA **Safety** manual.

The fire department can obtain insurance to protect itself from liability suits. There are various types of insurance coverage available. The department should check with knowledgeable insurance representatives concerning coverage for their department. Fire department personnel who purchase insurance MUST make certain they understand exactly what the policy covers. Finding out the policy does not cover something after it has happened does not help the department or the instructor. When purchasing liability insurance, medical malpractice coverage should be an important consideration. Errors and Omissions insurance is professional liability insurance. Its purpose is protecting the fire department, including fire department personnel, from financial loss resulting from liability lawsuits.

INSURANCE

Fire departments should have a risk management program that evaluates potential problems, stresses safety, and provides insurance. The purpose of insurance is to provide protection from potential losses. Insurance, however, does not replace safety. Fire departments should have Worker's Compensation Insurance for employees. Worker's Compensation Insurance will protect the department from claims arising from injury to employees.

The department may also carry liability insurance. Liability insurance provides protection from the results of negligent acts. Property insurance protects property such as buildings. Departments will also want to insure their vehicles. Insurance policies vary widely in their coverage and scope. Insurance laws also vary from state to state; departments should check their state regulations. Because of the complexity of insurance and the various policies available to fire departments and individuals, fire departments should research their state laws and consult with appropriate personnel.

Personal Insurance

Instructors may want to consider personal liability insurance coverage in addition to coverage through the fire department. Personal insurance coverage is purchased by the instructor, beyond that provided by the fire department. Instructors should check their departmental coverage and then determine if they need additional coverage. The instructor should understand exactly what such a policy covers and under what conditions it is effective.

EQUAL EMPLOYMENT OPPORTUNITY AND AFFIRMATIVE ACTION

Equal employment opportunity (EEO) is a personnel management responsibility to be sensitive to the social, economic, and political needs of a jurisdiction or labor market. Under the Equal Employment Opportunity Act of 1972, government agencies and private industries with federal government contracts are required to develop policies to assure equal access to jobs based on the applicant's ability to perform the job. This concept does not imply that all applicants are equal; it does imply that they have an equal chance to qualify. Thus, the applicant must be systematically selected and promoted on job-related criteria, and not excluded because of gender, race or ethnic categories, disability, or religious preference (Figure 3.7).

Figure 3.7 The Equal Employment Opportunity Act helps ensure that applicants are hired on job-related criteria and are not excluded from consideration due to gender, race, ethnic background, disability, or religious preference.

As a member of the human resource development team, training officers and instructors are often called upon to play a significant role in personnel selection and promotion. In many cases, the training division does all the promotional activities for the department. Fire service instructors should be aware of the guidelines and requirements for promotions. The department should be consistent in the way it promotes individuals and should have set criteria for promotions. The fire service instructor should take the responsibility of promoting individuals seriously since this situation makes them potentially liable for numerous personnel management errors.

Many serious problems in today's fire service involve personnel selection. Public safety personnel have often been selected using no specific criteria or criteria not related to the job duties and tasks. As an example, some organizations have had arbitrarily determined height requirements that have eliminated some potentially qualified ethnic groups and female candidates. If standards are unrealistically high compared to local requirements, modifying the standards will not necessarily make them excessively low.

Personality and other standardized tests have had an impact on the selection process. These tests should be used with extreme caution. They should be administered by a qualified professional, such as a licensed psychometrist, in accordance with the *Uniform Guidelines on Employee Selection Procedures* and the principles and practices recognized by the American Psychological Association.

Developing valid job selection criteria is not enough; the selection process itself must also be valid. A common administrative error is the failure of management to train raters. Many fire service agencies have been faced with selecting a few people from a pool of over 1,000 applicants. Such a situation, if not managed well, can become administratively overwhelming. Raters must be properly prepared to rate candidates. Problems are created from one or more of the following situations:

- Raters are not carefully selected.

- Raters vary in their interpretation of the rating scale.

- Rater teams vary in their qualifications.

- Rating scales are poorly defined, using such non-measurable terms as "average," "above average," and "excellent."

- Fatigue causes some raters to grade differently as the day gets later or from one day to the next.

In order to avoid these and other related problems, well-qualified personnel managers thoroughly versed in personnel selection and promotion should be consulted. It may also become necessary to consult with a labor attorney, an industrial/organizational psychologist, and/or a psychometrist. Finally, every member of the selection and training team must be thoroughly trained to ensure their reliability.

A number of resources are readily available to help provide training officers and instructors with a working knowledge of EEO. One of the most informative is Bulletin Number FSTB-402 *Achieving Job-Related Selection For Entry-Level Police Officers and Firefighters* developed by the U.S. Office of Personnel Management and distributed by the International Association of Fire Chiefs.

Fire departments that interview potential employees or have them fill out application forms should realize the legal factors involved. In many instances, it is against state or federal law to ask certain types of questions. For example, it is not appropriate to ask a candidate's religious preference. Only information pertinent to the job of fire fighting may be asked.

Closely related to EEO is affirmative action. Affirmative action employment programs are designed to make special effort to identify, hire, and promote special populations where the current labor force in a jurisdiction or labor market is not representative of the overall population. Thus, affirmative action may become an integral part of an organization's EEO program (Figure 3.8).

Figure 3.8 The goal of Affirmative Action programs is to have a work force that reflects at every job level the sexual and ethnic composition of the surrounding area. *Courtesy of Columbia Fire Department (Missouri).*

The purpose of affirmative action is to correct past inequities to special populations. Special persons are individuals who may require such interventions as focused training, counseling, and/or experiences to become qualified. Some organizations have developed intervention programs to improve physical strength and coordination, mechanical knowledge and skills, and study habits. Structured pre-employment experiences have been used to further develop qualifications.

A well-conceived affirmative action program requires a total commitment at all levels of an organization. Management may assign training personnel the responsibility to provide not only the intervention strategies, but also to give an orientation for personnel in the organization.

Affirmative action is a corrective process, and an organization may be held responsible for meeting employment quotas with specific milestones.

Drug Testing

Drug testing has become a sensitive legal issue in today's fire service. Drug testing is used when hiring employees, promoting employees, and for random checks. When administering drug tests, fire department officials should make certain proper procedures are followed. The department's drug testing policy should be in writing, and there should be no exceptions. Departments may want to consult with legal counsel regarding their drug testing policy or before initiating a drug testing procedure. The fire service demands that employees be alert mentally and physically. Drug testing is one method of ensuring employees are fit for duty but it must be done with considerations regarding legal factors.

COPYRIGHT

The Copyright Act of 1976, which became effective on January 1, 1978 is designed to protect the competitive advantage developed by an individual or organization as a result of their creativity. Infringement may result in litigation and can carry stiff penalties.

The law has not been fully tested in the courts, and thus, it is difficult to describe the potentially great variety of instances that may violate the copyright act. This section, however, does provide some general guidelines.

For a copyright to be valid, it must be filed with the United States Copyright Office. A valid copyright gives the owner sole rights to:

- Reproduce the work.
- Prepare derivative work.
- Distribute work.
- Perform work publicly.
- Display individual images.

There are numerous situations that clearly violate copyright law. Some obvious ones include photocopying workbooks and duplicating video productions without permission of the copyright owner. One may also violate copyright law by using copyrighted music for background in a slide-tape presentation or by playing a copyrighted film over a cable television without the owner's permission. To avoid infringement, one should contact the copyright owner, stating the name of the work and under what conditions it will be used. Once written approval has been received, credit for use of the work must be given.

Under certain conditions, "fair use" of copyrighted material is permitted. The general rule is that permission for use is not required for criticism, comment, scholarship, and research. In testing for "fair use," consideration must be given to:

- The nature of the copyrighted work
- The substantial nature of the portion used
- The effect on the potential market, or loss of profit

Thus, the purpose and character of use must be viewed in its totality and reasonableness. Even under "fair use" conditions, credit must be given to the author. If, as an example, an instructor was to copy a chart from a text to be handed out in a class, the author of text must be noted.

OBTAIN PERMISSION FOR PICTURES

When a fire department takes photographs or shoots film, permission should be received from the individuals being filmed or the owners of the property being filmed. Individuals generally have the right to control the use of pictures of themselves or their property. The fire department could be sued for invasion of privacy or libel if appropriate permission is not obtained. The extent of the fire department's legal responsibilities will depend on different factors such as where the film was shot (for example, was it shot in a public or private setting). Because of the serious consequences, both legal and professional, the department should always play it safe. Fire departments should obtain permission from individuals by having them sign a release form (Figure 3.9). It is best if the fire department obtains both prior and subsequent consent. This means that the department should obtain permission before taking pictures or film and also obtain permission after the pictures or film are developed and the individuals have had a chance to review the pictures or film.

> **I** grant without recourse, total and complete authorization to the (Name of Fire Department) for all photographs, negatives, proofs, or slides which have been taken of me, or of any property or materials owned by me this day, for instructional, educational, or any other purpose without compensation or renumeration to me, and that all photographs, negatives, proofs, or slides shall remain the property of (Name of Fire Department).
>
> Signature_____
> Address _____ Phone_____
> City _____ State _____ Zip _____
> Signature of Parent or Guardian if Minor
>
> _____
> Signature of Witness _____ Date _____

Figure 3.9 A release form should be signed by individuals who are being filmed or whose property is being filmed.

SUMMARY

Fire service instructors must be concerned with the legal aspects involved in the fire service. The ever-increasing number of lawsuits makes it imperative that fire service instructors be aware of laws within their jurisdiction that, in essence, could be used against them or the department. Because of the seriousness of litigation, fire service instructors must thoroughly understand their responsibilities and the direct impact legal action could have on their position and the department.

Liability is a broad term that refers to a person's or organization's responsibility in the law. If a plaintiff has the four elements for tort prosecution, then the fire service instructor personally and/or the municipality may be held liable. Fire service instructors should make every attempt to minimize the risk of a liability action.

The administrative and management duties of the fire service instructor also require that the instructor consider legal implications when performing these duties. Fire service instructors should be aware of the legal implications surrounding equal employment opportunity, affirmative action, copyright, and right of privacy.

SUPPLEMENTAL READINGS

Anderson, Betty R., and Martha P. Rogers. *Personnel Testing and Equal Employment Opportunity.* Washington, D.C.: Equal Employment Opportunity Commission, 1970.

Bahme, Charles W. *Fire Service & The Law.* Boston: National Fire Protection Association, 1976.

Berry, Dennis J. *Fire Litigation Handbook.* Quincy: National Fire Protection Association, 1984.

Jenaway, William F. *Fire Department Loss Control.* Ashland Massachusetts: International Society of Fire Serv ice Instructors, 1987.

Kigin, Denis J. *Teacher Liability in School Shop Accidents.* Ann Arbor, Michigan: Prakken Publications, 1973.

Rosenbauer, D.L. *Introduction to Fire Protection Law.* Boston: National Fire Protection Association, 1978.

Valente, William D. *Law in the Schools.* Columbus, Ohio: Charles E. Merrill Publishing Co., 1980.

The Psychology
of Learning

This chapter provides information that addresses performance objectives in NFPA 1401, *Fire Service Instructor Professional Qualifications* (1987), particularly those referenced in the following sections:

NFPA 1401

Fire Service Instructor

3-1.1 (a)(b)(c)(d)(i)(j)(k)

3-1.2 (c)(f)(m)(o)(s)

3-4.1 (a)(b)(c)(d)(f)

3-4.2

3-5

3-9.1 (f)

3-9.2

3-10.4 (a)(b)

Chapter 4

The Psychology of Learning

An instructor who understands the psychology and basic principles of learning will better understand the effectiveness of various teaching methods. Questions that confront the instructor are: What is learning? How does learning take place? What psychological factors are involved? How does a person receive new ideas?

As the instructor learns to use various techniques for the presentation of course material, it is necessary to successfully apply the proper techniques and to understand how and why a student learns.

> **Learning is a *relatively permanent* change in behavior that occurs as a result of acquiring new information, skills, or attitudes from or through an experience. Usually, it is reinforced through practice; therefore, frequency and intensity of experience enhance learning.**

Learning is related to, but different from, maturation, growth, development, or thinking. Thinking can lead to learning, but it is not the same. Learning is a result of experience. It is something that occurs within a person, which is not directly observable, but its effects can be seen through changes in behavior. Learning can be purposeful, like that in school or firefighter training. It can be incidental, such as learning that a distant fire is burning based on the observation of smoke on the horizon.

Learning is most likely to occur under conditions of focused attention and deliberate effort (Figure 4.1). When individuals want to learn, are motivated enough to focus on learning, and *continue to practice* the new behavior, they learn rapidly and efficiently.

Figure 4.1 Effective learning in the classroom takes concentration and a desire to learn.

TYPES OF LEARNING

Learning has been classified in many different ways. Three classifications are based upon the type of learning involved. They are cognitive (knowledge) learning, psychomotor (skills) learning, and affective (attitude) learning. Each type of learning progresses from less difficult to more difficult levels. These classifications are useful, since certain learning principles apply more directly to one kind of learning than to others.

Cognitive (Knowledge) Learning

The most commonly understood learning domain is the cognitive domain. The learning that results from instruction in the cognitive domain deals with recall or recognition of knowledge and the development of intellectual abilities and skills. The cognitive domain is addressed in a technical information presentation, generally using the lecture teaching method.

Each level of learning in the domains builds upon the previous level and therefore is progressive in nature (Figure 4.2). The levels of learning in the cognitive domain are shown below and were developed by Bloom and his associates.

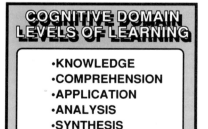

Figure 4.2 Each level of learning in the cognitive domain builds upon the previous level.

- Knowledge — Recalling and recognizing information.
- Comprehension — Understanding the meaning of information.
- Application — Using information learned in specific situations.
- Analysis — Breaking information into parts to understand the whole.

- Synthesis — Integrating the parts to invent a new whole.
- Evaluation — Using standards and criteria to judge the value of the information.

Understanding the cognitive domain enables the instructor to plan more effective instruction. Large units of instruction can be broken into smaller pieces so that they build upon one another until the whole unit is learned. Some examples of cognitive learning are:

- Learning terms, facts, procedures, or principles.
- Being able to translate and explain facts and principles.
- Being able to apply facts and principles to a new situation.

Examples of higher levels of cognitive learning range from being able to analyze a situation, to formulate a plan based on the analysis, to evaluating someone else's plan.

Psychomotor (Skill) Learning

The most commonly used domain of learning in the fire service is the psychomotor domain. Psychomotor learning refers to the ability to physically manipulate an object or move the body to accomplish a task. "Psychomotor" refers to the skills involving the senses and the brain as well as the muscles. Examples of this kind of learning may progress from recognizing the steps of a task and the tools that must be used, to trying the skill with guidance from the instructor, to practicing it under supervision, to conducting the skill independently.

Just as in the cognitive domain, each psychomotor level is progressive, building one upon the other (Figure 4.3). The levels of learning in this domain are shown below.

- Observation — Witnessing a motor activity.
- Imitation — Copying a motor activity step-by-step.
- Adaption — Modifying and personalizing a motor activity.
- Performance — Perfecting the activity until the steps become habitual.
- Perfection — Improving performance until it is flawless and artful.

Affective (Attitude) Learning

The least commonly used or understood domain of learning is the affective domain. This domain is the feeling or attitude-related aspect of instruction. The affective domain is rarely taught as specific subject matter like the cognitive and psychomotor domains. Affective learning relates to the awareness, attitudes, interests, appreciations, and values of a student. Examples of this type of learning include listening attentively in

PSYCHOMOTOR DOMAIN LEVELS OF LEARNING

- •OBSERVATION
- •IMITATION
- •ADAPTION
- •PERFORMANCE
- •PERFECTION

Figure 4.3 The levels of learning in the psychomotor domain are progressive in nature.

class, willingly participating in class, appreciating something enough to do it outside of class, and making it a basic part of one's life-style. These attitudes and values are essential to learning.

Just as the cognitive and psychomotor domains are progressive in nature, the affective domain progresses from simple awareness to acceptance to internalization of and acting out chosen attitudes (Figure 4.4). The levels of learning in this domain are listed as follows:

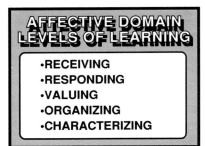

AFFECTIVE DOMAIN LEVELS OF LEARNING
- •RECEIVING
- •RESPONDING
- •VALUING
- •ORGANIZING
- •CHARACTERIZING

Figure 4.4 In the affective domain, the levels of learning also build upon one another.

- Receiving — Becoming aware of concept.

- Responding — Indicating that the concept has been received.

- Valuing — Internalizing and committing to some position.

- Organizing — Internalizing and adjusting among values.

- Characterizing — Adopting and personalizing the concept or value.

The value system of an individual has controlled his/her behavior for a sufficient amount of time for that individual to develop a characteristic "life style."

The instructor must realize that learning in the affective domain takes time to achieve and is not readily observable. This differs from learning in the cognitive and psychomotor domains in which learning may be observable within the classroom. Affective learning can be inferred from other behavior. For example, a trainee may value safety. The observable behavior of a trainee, which demonstrates a safety consciousness, could be reporting hazards, wearing safety equipment, or following rules without being reminded.

Although the affective domain is difficult to measure, it should not be overlooked by the instructor. In the affective domain, it is acceptable to look at indicators rather than specific measures to determine learning.

Clues For Recognizing The Types Of Learning

Often there are clues for recognizing the different types of learning. One clue is *when* the instructor expects to see the results. When the instructor implies that tasks will be learned "upon completion of the lesson, course, or certification series," it indicates that the learning will be either knowledge or skill learning. However, if the instructor implies that learning may not be seen until the student returns to the job or until an attitude has changed, then attitude learning is occurring.

Another clue is *where* the training will occur. If training is in the classroom, knowledge is being taught and learned. If training is at the training facility, skills are being taught and learned. Attitudes can be taught in the classroom or on the fire training ground.

A final clue is *how* the learning will be tested. If learning is tested by a performance test, it is typically skill learning. A performance test, however, can also test attitudes, such as safety. If learning is measured by a written test, it is usually cognitive learning. Attitudes can be tested by observation during a performance test or through paper and pencil tests that help the student project attitudes by responding to situations.

STUDENT MOTIVATION

A student must have a desire or need to learn before there can be comprehension. This desire or need is called **motivation**. Motivation has been described as a process in which energy produced by needs is expended to achieve goals. If the student is motivated, he or she will be better able to comprehend, retain, and use the information received. Motivation is created from within the student, not by the instructor.

Motivation is an important factor in instruction. Instructors can make a conscious effort to draw out student motivation. Motivation, on the part of the student, leads to persistent effort toward the learning objective. When people want to learn and they understand the reason for what they are learning, they generally learn at a more accelerated rate. For this reason, it is necessary for the instructor to understand the basic drives that motivate people (Figure 4.5).

Figure 4.5 Motivation, from within the student, accelerates learning.

People tend to have certain wants or needs that they strive to satisfy. **Trying to satisfy these needs creates motivation.** Abraham Maslow has identified these basic human needs and placed them in the Hierarchy of Human Needs. Listed in ascending order of importance, these needs are physiological, security, social, self-esteem, and self-actualization. Instructors should be aware of these needs and how they affect the student's behavior during instruction (Figure 4.6).

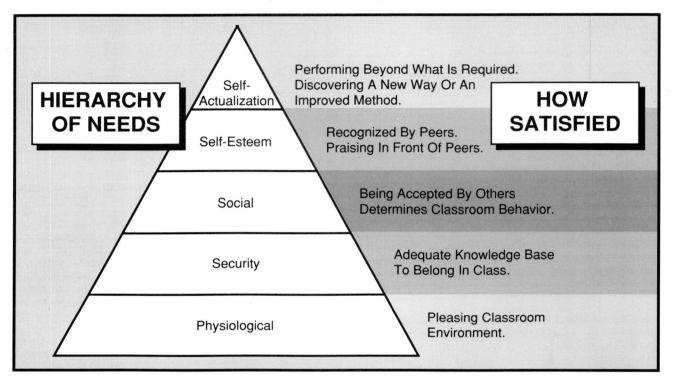

HIERARCHY OF NEEDS

HOW SATISFIED

Self-Actualization — Performing Beyond What Is Required. Discovering A New Way Or An Improved Method.

Self-Esteem — Recognized By Peers. Praising In Front Of Peers.

Social — Being Accepted By Others Determines Classroom Behavior.

Security — Adequate Knowledge Base To Belong In Class.

Physiological — Pleasing Classroom Environment.

Figure 4.6 Abraham Maslow identified basic human needs and placed them in a sequential hierarchy.

Physiological needs can be satisfied by placing the student in a pleasing classroom environment. Fatigue and irritating distractions should be avoided. Suitable seating arrangements, adequate lighting, and temperature control are necessary.

The need for security can be fulfilled by making sure the student is suited for the class. The instructor may need to reassure the student that he or she has an adequate knowledge base to keep up with other students and be successful in the class. If necessary, the instructor can provide special tutoring to help a student better comprehend the material.

A student will probably try hardest to satisfy social needs. The natural human tendency to have a feeling of belonging is important; therefore, students will strive to associate with and be accepted by their peers. This tendency will govern the students' activity in class and will probably explain their classroom behavior.

Self-esteem needs can be satisfied by recognizing a student's competence or knowledge in front of the student's peers. A student who has the opportunity to demonstrate satisfactory performance of a difficult task will be thought of highly by other students.

Satisfying the self-actualization, or self-fulfillment, need is sometimes difficult. The effort needed to satisfy lower level needs diverts a person's energies and the need for self-fulfillment remains relatively dormant. However, some people do fulfill their potential. By becoming successful in some endeavor beyond what is required, people may satisfy this need.

Although motivation originates from within the student, the instructor can help the student become motivated (Figure 4.7). If there is a block that prevents the student from attaining chosen goals, the reaction will be either a fight to drive through the block or a tendency to withdraw and give up. It is the instructor's responsibility to eliminate or minimize such blocks.

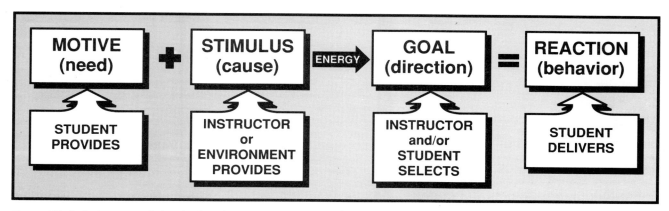

Figure 4.7 An instructor can help stimulate the students' interest so they become motivated.

LEARNING INFLUENCES

There are many influences that affect a person's ability to learn. Instructors should know and be able to apply these in teaching. Influences or conditions of learning are the basis for instructional methods and aids. The instructor can exert some control to facilitate the conditions that influence learning. Therefore, the effective instructor attempts to design learning activities with the following factors in mind.

Instructor Attitudes

The instructor's attitudes either positively or negatively influence learning. Instructors' attitudes expressed through their behaviors positively influence learning when they:

- Express the belief that anyone can learn a new skill.
- Reduce stress and frustration in learning experiences.
- Accept individual differences.

- Encourage freedom of expression.
- Create a pleasant classroom environment.
- Promote success.
- Give recognition for even the smallest success.

Goals, Objectives, And Standards

A sense of purpose or direction is often the force that motivates a student. There must be some identification with the direction of the class. Setting goals and behavioral objectives with the students will make identifying with the class direction easier for the student. Setting performance standards will provide the benchmarks to judge progress toward goals and objectives.

Relevance

Relevance is important to the student's ability to learn. If the material being presented directly relates to something the student is already doing or wants to do, it is much easier for learning to take place (Figure 4.8).

Figure 4.8 Learning will take place much easier when instruction is relevant to the job.

Preparation

Preparation for learning can enhance the learning experience. Introduction to the subject matter must include an explanation why the material is important and how it will be useful to the student. An overview of the material to be covered in the lesson mentally prepares the student for what is to come. Linking the subject matter to previously known information also aids the student in learning new subject matter.

Sequence

Learning is best accomplished by an organized, step-by-step procedure. If instruction is sequenced from the known to the unknown and from the general to the specific, it is easier to follow the instruction. So strong is the tendency to learn in an organized way that even when material is presented in a disorganized or relatively meaningless fashion, students tend to develop an organization of their own (Figure 4.9).

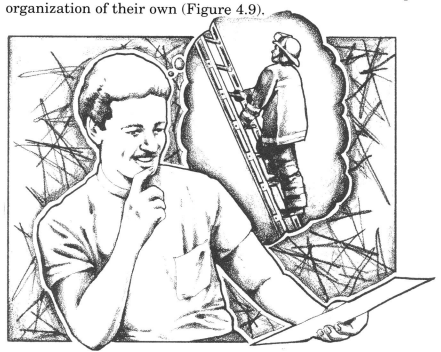

Figure 4.9 Students will organize information in order to help them reach their goals.

Participation

Learning is an active process. Action and involvement increase the probability of learning occurring. They also strengthen the retention of learning.

Practice And Repetition

The student retains learned material through practice. A person who sets out to do a task and accomplishes it successfully is better able to retain the knowledge needed for performing the task. The more often the knowledge is put to use, the better it will be retained. The student should practice new skills as soon as possible (Figure 4.10). If there is a large gap of time between learning and application, other learning in the interim may cause interference, resulting in the student forgetting what was learned earlier. New learning may also be distorted by previous learning. Difficulties may arise when a person who has learned to perform a task one way must learn to perform that task a new way. This person will often have more trouble adjusting to the new way than will someone without previous experience.

Feedback And Reinforcement

Generally, learners need immediate feedback and reinforcement. Feedback and reinforcement are needed early and often. Mistakes need to be corrected before learning is set. Successes need to be rewarded to encourage continued learning.

Figure 4.10 Practice helps the student better retain new knowledge.

Previous Experience

A person who has experience or knowledge in a particular area will be more receptive to new information related to that area than will a student who has had no such exposure. This is true because there is a transfer of knowledge whenever previous learning has an influence on new learning. Simply stated, the transfer of learning uses the student's experience. The instructor helps link new experiences with old and permits the student to gain insight or to solve a problem. This can be accomplished by the instructor continually encouraging students to think of alternative ways to question and apply knowledge.

NEGATIVE INFLUENCES ON LEARNING

Individual learners sometimes experience frustrations. This may result in using escape mechanisms. The person who stays frustrated develops habits to cope with these situations. Instructors must realize that these coping habits are symptoms of more serious problems. Realizing this, the instructor should look for the underlying causes. Once the causes are discovered, some of these frustrations may be eliminated (Figure 4.11).

FEAR OR WORRY
- Of the class situation
- Of failure
- Of ridicule
- About personal problems
- About family
- About health
- About money

DISCOMFORT
- Personal strength and stamina
- Eye strain
- Difficulty hearing
- Classroom too hot, stuffy, too cold
- Dangerous training conditions

POOR INSTRUCTION
- Class too advanced
- Class too simple
- Instructor unprepared
- No opportunity for participation
- No variety in presentation
- Class too large

Figure 4.11 Instructors must discover and eliminate student frustrations.

Some frustrations come from fear and worry. These may be related to the classroom, like fear of not knowing how to study, fear of ridicule by the instructor or classmates, or fear of failure. Other worries may stem from personal life, a sick family member, money matters, or home management concerns.

Many frustrations may arise from discomfort in the physical or class setting. Sitting or standing too long, not being able to see everything going on in class, and poor ventilation or temperature control can negatively impact learning. Unnecessarily dangerous working conditions or poor training ground organization can also hinder learning.

Other frustrations stem from boredom. If the student is not interested in the subject, if the instructor is lecturing too long, or if the student has little chance to try the new skill, then boredom may set in. Lack of any training aids or improper teaching methods can cause boredom and reduce learning. Success in any learning situation depends upon the individual and the material being learned. Instructors must realize this and try different methods to make the students interested.

Uncovering the negative influences on learning can help the instructor remove them, enhance the instruction, and increase the chance for learning. There are several factors that can have negative influences on learning.

Long Time Spans

Trying to learn too much during a long, continuous period may interfere with learning capacity. If an instructor teaches for one or two hours without a break, the information taught in the latter part of the course may not be thoroughly comprehended. Learning is best accomplished through short, intensive sessions, especially after lunch or on a hot afternoon. It is helpful to be alert to and read the body language of students.

Emotional Attitude

The emotional attitude of the student is vital. An individual will have difficulty learning if there is concern for real problems (a death in the family) or imaginary problems (an uncertain feeling about not being included in the group). The successful student overcomes negative emotional attitudes by maintaining the need or the desire to learn.

Learning Plateaus

Plateaus in learning are sometimes created by emotions, such as hate, fear, and boredom, that occupy the student's mind and interfere with concentration on the subject being learned. Plateaus can be compared to the landing in a flight of stairs. After students master the procedural steps necessary to perform the skill, they should be encouraged to practice until a desired skill level is attained. A student may become discouraged if he or she has not been able to practice a task enough to become proficient at it. The student may still have to think about each step or procedure as work continues. At this point, further progress might seem impossible, and the student may feel like giving up. Athletes often experience plateaus in developing their skills.

The good instructor informs students of these characteristics of learning and helps them recognize and overcome problems. The best solution to the problem usually is to continue practicing the procedure until it is thoroughly understood and becomes automatic. When plateaus continue to exist, the student may have formed improper habits or tried to learn something beyond individual ability. On the other hand, the instructor may have failed to provide proper assistance.

HOW WE LEARN

- 1% through TASTING
- 1 1/2% through TOUCHING
- 3 1/2% through SMELLING
- 11% through HEARING
- 83% through SEEING

Figure 4.12 The five senses are important to learning.

LEARNERS RETAIN

- 10% of what they READ
- 20% of what they HEAR
- 30% of what they SEE
- 50% of what they SEE and HEAR
- 70% of what they SAY
- 90% of what they SAY while they DO something

Figure 4.13 The best retention occurs when students explain what they are doing while they do it.

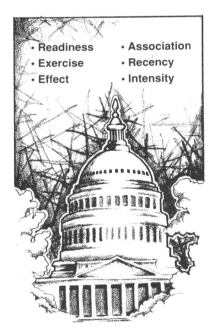

- Readiness
- Exercise
- Effect
- Association
- Recency
- Intensity

LAWS OF LEARNING

Figure 4.14 The six laws of learning.

For all practical purposes, the mind can be thought of as blank at the time of birth. Learning begins the moment the new life responds to the stimuli of the outside world. These messages stimulate the senses through which the mind can learn. The five senses are the windows through which the mind looks at the world. The following breakdown shows the relative importance of the senses for learning (Figure 4.12).

It has been shown that humans retain 10 percent of what is read, 20 percent of what is heard, 30 percent of what is seen, 50 percent of what is seen and heard, 70 percent of what is said, and 90 percent of what is said while doing (Figure 4.13). From this it can be concluded that the most effective mode of learning would include receiving a new idea and then, while performing a task, repeating what has been learned.

The instructor who fails to consider the physiological aspects of the learning process can only hope for partial success. The instructor must be concerned with how people learn, motivational factors, and retention. The instructor should also remember that each student is an individual and will not necessarily react in the same way.

LAWS OF LEARNING

Learning is a basic process of life. It is based upon certain recognized principles. To be successful, one must understand the laws that govern the learning process. Edward L. Thorndike's learning laws, the first three of which are given here, have withstood the test of time. The other learning laws were developed by contemporaries of Thorndike (Gutherie, Hull, and Miller), while building on his work (Figure 4.14).

Law Of Readiness

The law of readiness means a person can learn when physically and mentally adjusted (ready) to receive stimuli (instruction). Visualize the concept by imagining a cat ready to pounce on a bird or a child struggling to reach a toy. This readiness to learn is evident in a class where students show high interest and anticipation of the activities to be carried out during a lesson.

Law Of Exercise

The law of exercise stresses the idea that repetition is basic to the development of adequate responses. Certainly, no one ever becomes proficient at a skill without performing the operation over and over. The amount of repetition required will vary from person to person. Learning is always based on activity, which requires some kind of exercise involving both mind and body. It should be pointed out, however, that practice does not always "make perfect." Mere repetition may be dull and meaningless if the student cannot see and appreciate the reason for it. Repetition is next to useless without the essential elements of interest, meaning, and goal.

Law Of Effect

Learning will always be more effective when a feeling of satisfaction, pleasantness, or reward accompanies or is a result of the learning process. This is not to say that learning is always painless. People often learn worthwhile lessons by "suffering the consequences" of their actions. However, if the goal in view is desirable and will satisfy a need or desire, an individual is willing to suffer many setbacks on the road to success. Another factor influencing the effect of learning is the use of either praise or blame as a learning tool. Research shows that praise is more effective than blame in motivating students. In other words, although punishment may be necessary at times, reward is more effective.

Law Of Association

When the mind compares a new idea with something already known, it is using association. This means that it is easier to learn by relating new information to similar information from past experiences. Making associations is a great aid to learning. Instructors should, therefore, encourage associations and also try to help students make connections with previously learned material.

Law Of Recency

Reviews, warm-ups, and make-up exercises are all based on the principle that the more recent the exercise, the more effective the performance. It is difficult to separate this principle from the idea that exercise and practice are vital for learning. The principle of recency simply means that practicing a skill just before using it will ensure a more effective performance.

Law Of Intensity

The principle of intensity says that if the stimulus (experience) is real, there is more likely to be a change in behavior (learning). This does not necessarily suggest stimuli that are pleasant, rewarding, or satisfying. An example of this might be a demonstration of using rescue equipment for vehicle extrication, an experience that may be unpleasant, but is more apt to be remembered than a talk. In other words, an actual explosion is more effective than just the word "Bang!"

TRIAL AND ERROR

One way that learning occurs is by trial and error. This is generally considered to be the least efficient method. It can be time consuming, costly, and dangerous. It is also unavoidable.

The child, for example, would learn very little if it were not for trial and error attempts, just as the research scientist reaches a point where the next step must be "try it and see what happens." The latter is a calculated trial and error.

A way of learning that takes advantage of all the records of the trial and error of the past might be called guidance. This is illustrated by an instructor/student relationship in which the instructor's experience is passed on to the student, thus eliminating the need for the student to experiment.

When students are able to call on past experience to help solve new problems, it might be said that they are learning by analysis. This might be called trial and success, because experience suggests the right way to do the new job.

TRANSFER OF LEARNING

Transfer of learning refers to the opportunity to apply what has been learned in one situation to a new situation. Information, principles, and procedures are taught in one setting. The student generalizes what has been learned and applies it to a new problem. Transfer is not automatic for the student unless the instructor structures learning exercises, activities, and discussions that allow for transfer of learning to occur. *Hands-on, manipulative exercises are the single most effective means of transferring learning for the firefighter.* Group exercises and discussions are effective for transferring learning in areas such as officer and instructor training.

COMPETENCY-BASED LEARNING (CBL)

Training intended to create or improve professional competency must be based on performance, not content. The emphasis is on what the learner will be able to do, not what the teacher will teach. This is known as competency-based learning (Figure 4.15).

Figure 4.15 Competency-based learning focuses on what the learner will be able to do.

Competency-based learning grows out of identified and verified competencies of the profession. Competencies are inherent requirements of the job. Absolute standards or criteria of performance are based on the competencies of the profession. They are identified through a job analysis. One example of a standard of performance is the National Fire Protection Association Professional Qualifications.

Once standards of a profession or job are set, course goals and lesson objectives can be written. Competency-based instruction occurs, with the end result being competency-based learning:

- Competencies exist.
- Job analysis occurs.
- Standards are set.
- Course goals are written.
- Lesson objectives are written.
- Competency-based instruction occurs.
- Competency-based learning results.

Other terms are used to refer to instructional or evaluation approaches that use competencies as their foundation. Some common terms are criterion-based instruction, criterion-referenced testing, and performance-based instruction.

Mastery Learning

A critical element of competency-based learning is the fact that the outcomes are to be expressed in minimum mastery levels for each competency. Instead of averaging the scores throughout the course, each specific competency must be attained to a predefined level. With the mastery concept goes the responsibility of giving students adequate time to master each competency and to work at their own pace.

The emphasis of testing is on performance, not on time. In traditional learning, the time allowed to learn is the same for all students, but the achievement level varies among students. In mastery learning, the time varies for different students, but the achievement level is the same.

Behavioral objectives, instruction, and evaluation are tailored to performance, not just knowledge. Since mastery is a basic idea in competency-based instruction, individually paced learning is the key to achieving mastery. This type of learning takes into account the individual's needs when designing the learning environment. This means giving some students a longer time to learn. It means supplying immediate feedback on learning progress so the student can move through the activity more efficiently. It means offering a variety of realistic learning experiences that the student can relate to. It means that a classroom/group learning environment can be used as well as a one-on-one approach.

At testing time, competency-based instruction holds both the instructor and the student accountable for mastery.

Traditional Vs. Competency-Based Learning

The basic differences between traditional and competency-based learning are shown in Table 4.1.

TABLE 4.1
CURRICULUM STRUCTURES

TRADITIONAL	COMPETENCY-BASED
Content-based	Competency-based
Time-based	Performance-based
Group-based	Individual-based
Group needs	Individual needs
Delayed, general feedback	Immediate, specific feedback
Textbook/workbook	Modules and multi-media
Instructor-dependent	Instructor-supported
General goals	Specific objectives
Norm-referenced assessment	Criterion-referenced assessment

INDIVIDUAL DIFFERENCES

Often, students are labeled as "problem students," when really they are expressing individual differences. There is a wide range of differences in students' abilities to learn.

Some people grasp ideas and master skills quickly, whereas, others learn more slowly. The instructor must gear the pace of learning to the student's rate of understanding; otherwise, the student may become discouraged. Good training often requires individual instruction. People also differ in their attitudes and emotional reactions. Some are timid or reserved; others tend to show off and try to impress the group.

These differences are symptoms. It is essential to understand some of the social and cultural implications that will affect many students in fire training.

Age

Everyone has probably heard a friend say, "I'm too old to learn new ways of doing things" or "I can't memorize things the way I did when I was a kid." Many adults make and believe statements of this type, but research has proven them wrong (Figure 4.16).

Contrary to popular opinion, the mind does not necessarily deteriorate with age. For example, in one research project a group

Figure 4.16 Research has proven that age does not hinder the ability to learn.

of 50-year-olds were given the same intelligence test they had taken 31 years before. They made high scores on every part of the test, with the exception of the mathematical reasoning section. In another study, men aged 20 to 83 took a course in world affairs at the University of Chicago. Follow-up studies showed the older students were more successful and continued to study the subject longer than did the younger students.

Students in the younger age brackets, however, are more apt to view authority, even that of the instructor, as suspect. They are more likely to ask "Why?" and to demand a definite answer. Knowing why it is important to study a topic is a handy tool for the instructor in motivating students, as well as for answering "Why?".

Adults have many personal characteristics that instructors can use to advantage in the classroom. These characteristics are as follows:

ADULTS HAVE MANY LIFE EXPERIENCES. Implications: Discussion techniques are useful for establishing relationships between the past experience of the learner and the subject to be taught.

ADULTS ARE HIGHLY MOTIVATED TO LEARN. Implications: Learning experiences must have immediate usefulness because many adults are concerned with personal achievement, satisfaction, and self-fulfillment.

ADULTS HAVE MANY COMPETING DEMANDS ON THEIR TIME. Implications: If student needs are not met, other activities will take higher priority. Students should be involved in planning and goal setting to make the course of instruction efficient.

ADULTS MAY LACK CONFIDENCE IN THEIR ABILITY TO LEARN. Implications: Learning experiences must satisfy needs and students must feel accomplishment at the end of each class.

ADULTS VARY MORE FROM EACH OTHER THAN DO CHILDREN. Implications: Meeting individual goals and providing varied activities and materials is necessary.

ADULTS LEARN BEST WHEN:

- They actively participate in setting learning goals.
- There is an appropriate learning climate (facilities, informal atmosphere)
- Learning is problem centered.
- They set their own pace.
- They receive feedback about their progress.

Subcultures

A major shift in social structure is taking place because of the rise in numbers and power of ethnic groups. Each group brings to the classroom their cultural behavior, attitudes, and values. The instructor needs special skills to use and enhance this diversity for the benefit of the learning environment. The instructor also must have a nonjudgmental attitude to realize that each subculture has value.

Educational Experience

A student's previous educational experience will greatly influence attitudes, confidence, and the ability to handle new learning experiences. The educational experience encompasses educational level, literacy level, and learning disabilities.

EDUCATIONAL AND LITERACY LEVELS

Educational levels and literacy levels have usually been considered the same. However, the educational level is the number of years spent in school, while the literacy level is the level at which students can read and write.

If a firefighter has completed high school, then it is assumed that the literacy level would be 12th grade. This assumption no longer holds true. In some departments, instructors are finding that the educational level of entering firefighters is higher than it has been in past years, but that the literacy level has declined.

How should instructors handle this low literacy level? The instructor's primary thrust should be through careful development of lesson plans and presentation methods. Instructors should use visual and training aids and avoid long lectures. Written materials, such as texts, tests, and handouts, should be written in short sentences and paragraphs. The format for written material should be double-spaced when possible, contain directional headings and wide margins, and be printed in type large enough so that it is easily read. Vocabulary used in the development of lesson plans and in the classroom should be as simple as possible. Provide glossaries of technical terms. Keep directions on tests and equipment "short, simple, and to the point." Many of these remedies are treated in greater detail in subsequent chapters on developing lesson plans and tests.

LEARNING DISABILITIES

Learning disabilities, of course, complicate the learning process. Because our educational system has ways of identifying disabilities, we are more aware of them than in the past. Almost everyone has some degree of a learning disability, and everyone finds unique ways of compensating for that disability. However, there are those whose disability is a major stumbling block to learning. Extra patience, understanding, and assistance are required of the instructor to make learning as productive as possible. Three primary means of helping the student with learning disabilities are tutoring one-on-one, developing individualized instruction, and providing feedback on progress.

INDIVIDUAL LEARNERS

Each class is made up of individuals. Some individuals in a class require more attention from the instructor than usual. Those demanding extra attention take time and effort away from the total class. Therefore, the instructor needs to recognize these individuals and know how to deal with them.

The ability of the student affects the amount of instructor attention required. Both gifted and slow learners require extra instructor attention. Level of classroom activity defines another group of learners. Nondisruptive students usually do not participate in class, but do not disturb it. Disruptive students actively jeopardize classroom management. Each demands extra effort from the instructor.

Gifted Learners

The gifted student is usually able to accomplish more than is expected of the average student. This type of individual may learn satisfactorily without much supervision. Gifted students are often ahead of other class members and are an asset to the class if the instructor makes proper use of their abilities. One way to handle this above-average group is to give them challenging assignments. It is not advisable to permit gifted students to be idle or to spend time on assignments that are below their level of ability.

Slow Learners

An instructor will identify the slow learners shortly after starting to work with the group, but should not jump to conclusions. After all, students may be slow to comprehend because the instructor is inefficient. For students who truly are slow learners, it may be necessary to arrange private conferences, special assignments, extra study, or individual instruction. The subject matter, training, or training methods used by the instructor may need to be reworked or revised, after which the individual may be able to keep up with the class. Some students may have difficulty grasping information and skills in one category, yet may excel in other areas. Therefore, assignments and lesson plans should be developed with the student's ability in mind. Instructors should always praise the accomplishments of the students (Figure 4.17).

Nondisruptive Learners
TIMID SOULS

The hesitant, tongue-tied individual is a shy, timid soul who is afraid to utter a word during class discussion. Fear keeps this person silent, although the student often has much to offer. Learning is most likely occurring, but in a passive rather than active way.

The considerate instructor should at first avoid calling on the timid person to recite or discuss topics before the group. This type of person can be encouraged to participate more freely if the atmosphere of the class is informal. Visiting during break may help this student feel more at ease in the classroom. The instructor can help the timid person overcome shyness and take a more active part in the class through simple questions and discussions. When a student is gradually led in this manner, timidity will eventually be overcome (Figure 4.18).

Figure 4.17 An instructor must realize that there are many reasons why a student may be a slow learner.

Figure 4.18 Fear keeps the shy, timid student silent in class discussions although he or she may have much to offer.

DAYDREAMERS/UNINTERESTED LEARNERS

Students may be normally above average in ability, but because of circumstances, such as uninteresting subject matter, use of unfamiliar terms, boredom with the instructor, and lingering too long on one or more points in the subject, the students may drift mentally. Daydreaming may be indicated when the student begins gazing around the room, doodling, or thumbing through materials not directly related to the subject. This individual's attention can sometimes be brought back to the class by direct questions or more active involvement (Figure 4.19).

The uninterested student displays very little energy and interest and has a low learning and accomplishment rate. When someone in the class seems to have these characteristics, obtain permission to secure records on the student's level of ability. Lack of interest is not natural and the cause should be found. This individual may have poor health, or be nervous, worried, or frightened. The student may lack the background for the training or the training may be personally distasteful.

If the class meetings are going to continue long enough, there are several things that can be done to convert this type of person. The instructor should first attempt to check on the health and living habits of the student. Simple personal questions will usually bring out the facts. The instructor may remind the student that there is a responsibility to learn, then try to inspire the person to greater ambition. The instructor should also try to shift or change the instruction so the person may become interested, but the shift should not be to the detriment of the other students (Figure 4.20).

Disruptive Learners

TROUBLEMAKERS

A successful fire service instructor cannot afford to tolerate inattention or troublemaking of any kind. When an inattentive, troublesome person is in a group, that person will distract the other students and prevent learning. The instructor should talk with the troublemaker, explaining the seriousness of that person's actions. The instructor should have the troublemaker evaluate why he or she is in the class, and explain that staying in the class will require a change in behavior. If the troublemaker continues to be a problem, the instructor should seriously consider expelling that person from the class. Although this may seem harsh, keeping a troublemaker in class is unfair to the other students and could prevent them from learning important material.

SIDETRACKERS AND STALLERS

A sidetracker or staller tries to divert the attention and interests of classmates from the lesson. One reason could be lack

Figure 4.19 An instructor can sometimes bring back the daydreamer's attention by asking direct questions.

Figure 4.20 Lack of interest in a student is not natural and if possible the cause should be found.

of class preparation by the student. A good instructor may keep this type of person in line by insisting that class time be consumed only by the lesson. The problem is usually solved by calling regularly on the class sidetracker so daily participation will be expected. Usually, a pointed word of caution about side-tracking is enough to call other class members' attention to the action of the offender. A personal word with the offender after class should eliminate this difficulty (Figure 4.21).

Another type of sidetracker is the talkative, aggressive, extroverted person who tries to monopolize the conversation. This type usually talks so much that no one else has an opportunity to talk. If a private personal appeal does not cure this person, suggest an extra assignment. While the offender is occupied, the other members of the group will have an opportunity to participate. There may also be members of a group who prefer to talk among themselves, rather than be attentive. Such a situation requires that the instructor recapture the attention of the group. Tell the offenders that special problems can be discussed after class when the whole group will not have to listen. If possible, talk to any problem individuals before class and enlist their help in class. Keep the purpose of the lesson in mind to direct class attention.

SHOW-OFFS

A show-off is a person who uses a group situation to perform acts of exhibitionism (Figure 4.22). There appears to be only one solution to this kind of problem: tell this individual in no uncertain terms that any disruption of the class will not be tolerated. The instructor should tell the student that the classroom is the

Figure 4.21 A sidetracker or staller tries to divert the attention of other students from a lesson so his or her lack of preparation will not become known.

Figure 4.22 A show-off uses the class setting to satisfy his or her need for attention.

wrong place for such behavior and that complete cooperation is expected. The timeliness with which the ultimatum is issued determines its effectiveness. When the student understands there is no alternative, the problem may be solved. If the problem persists, follow the L-E-A-S-T procedure recommended for effective class management.

INSTRUCTOR STRATEGIES

A few hints may prove helpful in handling problem students. First, get well acquainted with students in order to identify problems early. Second, draw upon the experience of other instructors. Ask them how they have dealt with similar problems or with the same student. Third, use the L-E-A-S-T method of progressive discipline.

THE L-E-A-S-T METHOD
OF PROGRESSIVE CLASS DISCIPLINE

L = At the first evidence of misbehavior, **"Leave it alone."** The behavior may have been an isolated occurrence.

E = If the behavior continues, make **eye contact** to convey dissatisfaction.

A = If the behavior continues, an **action** step is indicated. This is usually a comment stressing the importance of being attentive in class or directing a question to the problem student.

S = At this point the student is interrupting the class to the extent that you should **stop the class** and discuss the problem with the student. Taking a break is the most tactful way.

T = Assuming the discussion during the stop phase was not able to solve the problem, you should **terminate** this individual's classroom privilege (expel the student from the class) and take the appropriate disciplinary action.

Counseling

Successfully handling a conference is probably the most important aspect of dealing with problem students. It is also one of the most likely areas for instructor error. During the conference, the instructor and the student can explore problems of mutual concern and come to some understanding or agreement. The use of a conference should indicate that the instructor regards the student to be of potential value to the department or the class. The instructor should communicate this feeling to the student in positive terms. The student should be made to feel that the instructor has a sincere personal interest and views the student as a potential colleague. It should be obvious that the purpose of the conference is to help the student work out seemingly difficult problems that may interfere with individual ability (Figure 4.23).

Figure 4.23 The purpose of a conference is to work out problems that are interfering with student learning.

The student should be encouraged to explain any troublesome situation and to express feelings about it. The instructor should attempt to see the problems from the student's point of view, even while disagreeing. The instructor can point out, without being condescending, that others may have different points of view. In some instances, it is impossible to say who is right. It is, however, necessary for an organization to be operated on the basis of rules, regulations, assignments, and acceptance of responsibilities if it is to remain an organization. Care should be taken throughout the conference to maintain an atmosphere of unqualified, positive regard for the person.

Only after an instructor has exhausted every remedy should consideration be given to treating the problem as a disciplinary situation. Discipline should be done privately. In the majority of cases, problems result from a basic subconscious feeling of insecurity or inadequacy. In most instances, a sincere interest in the student will do more to solve a problem than will any treatment that intensifies feelings of inferiority or inadequacy.

Reprimanding

Attempts to force students into acceptable learning behavior patterns generally fail. One sure way to arouse resistance is to use force. An instructor must learn the techniques of stimulation and motivation, which are the result of positive reinforcement coupled with proper reprimanding. Good instructors only reprimand when absolutely necessary. When they do reprimand, they talk to the student in private, remain calm, establish a helpful atmosphere, know the facts and use them, begin the session with encouragement and praise for good work, suggest a constructive course of action, and criticize the mistake instead of the individual.

Helpful Hints

Several difficult situations may confront the instructor. Being prepared and having thought through situations ahead of time will help the instructor face situations successfully.

When the instructor is a peer to the students, it is critical to have adequate participation. The instructor can make assignments before the class so students can be prepared to get involved. Innovative teaching methods are supportive tools for involving students. Letting some students take a lead role in areas of specialty gets them involved. Pretesting to identify weaknesses allows the instructor to focus teaching time on skills for which there is a clear need. Finally, adequate preparation is essential for credibility with peers.

Handling cliques is sometimes trying for the instructor. A few suggestions can help to prepare for or prevent the problem. When convenient, meet with the supervisors of class members before the class to identify a possible problem. When anticipating cliques, divide the class by having students number off into groups and reseat themselves. Another solution is to prepare seat assignments before the class and ask everyone to sit in their assigned seat. Develop competition among and between groups, pairs, or individuals to break up or neutralize cliques.

It is critical for the instructor to "read" students and their body language. This can keep the instructor in touch with students' moods, needs, or frustrations. A quizzical look may indicate a lack of understanding and the hesitancy to ask a question (Figure 4.24). A slouched sitting position with arms crossed may indicate disagreement, disinterest, or lack of willingness to participate (Figure 4.25). A student nodding off may mean lack of interest or a sleepless night. In any case, the instructor needs to take action. A change of pace, a different teaching method, a break, or added enthusiasm by the instructor can help students overcome blocks that are keeping them from learning. In some cases, a brief visit at break with a student can give some insight to the instructor. It can also encourage the student to feel more at ease with the instructor and with the class.

All these tips are provided to ensure greater success for the inexperienced instructor. However, be assured that with time

Figure 4.24 The instructor should be aware of signals that may indicate a lack of understanding.

Figure 4.25 A student's body language can alert the instructor to the feelings of the student.

and experience class management skills will grow. The instructor will gain confidence and expertise in managing a class skillfully.

SUMMARY

The instructor who understands the factors that influence learning and the problems that individuals may have in learning will be a more effective teacher. Three types of learning are cognitive (knowledge) learning, affective (attitude) learning, and psychomotor (skills) learning. Each type of learning can be related to the fire service.

Students will be able to learn better if they are motivated. Motivation is a desire to satisfy needs. Abraham Maslow classified the basic human needs and formulated the Hierarchy of Human Needs. These needs are physiological, security, social, self-esteem, and self-actualization. Lower level needs must be met before an individual can focus on a higher level of needs.

In addition to motivation, there are many other factors that can either positively or negatively influence learning. Instructor attitudes, identification and involvement with the class, feedback and reinforcement, previous experience, and emotional attitude are just some of the factors that can influence learning.

Learning is a process that is based upon certain recognized principles. Edward L. Thorndike's laws of learning include the law of readiness, the law of exercise, and the law of effect. Learning that focuses on what the learner will be able to do is termed competency-based learning. The emphasis in competency-based learning is on performance and achieving minimum mastery requirements. This differs from the traditional concept of learning and is more applicable to the fire service.

Learning is also influenced by individual differences in students. Age, educational level, and personality are some individual factors that affect learning. The instructor should be able to deal with the differences in individual students including students who are gifted learners, slow learners, shy or timid learners, daydreamers or uninterested learners, sidetrackers or stallers, and show-offs. Students are individuals and the instructor must take that into account when teaching.

SUPPLEMENTAL READINGS

Bloom, B.S., ed., et al. *Taxonomy of Educational Objectives: Handbook 1, Cognitive Domain*. New York: David McKay Company, Inc., 1956.

Gagne, Robert M. *Conditions of Learning*. New York: Holt, Rinehart and Winston, 1977.

Gagne, Robert M., and Leslie J.Briggs. *Principles of Instructional Design*. New York: Holt, Rinehart and Winston, 1974.

Jones, Philip G., ed. *Adult Learning In Your Classroom: The Best of TRAINING Magazine's Strategies and Techniques for Managers and Trainers*. Minneapolis: Lakewood Books, 1982.

Knowles, Malcolm. *The Adult Learner: A Neglected Species*. Houston: Gulf Publishing Company, 1973.

Krathwohl, D.R., ed., et al. *Taxonomy of Educational Objectives: Handbook 2, Affective Domain*. New York: David McKay Company, Inc., 1964.

Mager, Robert F. *Developing Attitude Toward Learning*. 2nd ed. Belmont, California: Pitman Learning, Inc., 1984.

Simpson, E.J. "The Classification of Educational Objectives in the Psychomotor Domain." *The Psychomotor Domain*. Vol. 3. Washington: Gryphon House, 1972.

5

Planning
Instruction

This chapter provides information that addresses performance objectives in NFPA 1401, *Fire Service Instructor Professional Qualifications* (1987), particularly those referenced in the following sections:

NFPA 1401

Fire Service Instructor

3-1.2 (a)(b)(d)(g)(h)(j)(k)(p)

3-8

3-10.2

4-1 (a)(d)

4-2

4-3

4-4 (a)(b)(c)(f)(g)(h)(i)(j)(k)

4-5 (d)(e)

5-1 (a)(e)

5-3.1 (c)(d)(e)

Chapter 5

Planning Instruction

Excellence begins with curriculum. A well-designed curriculum is the foundation of a successful teaching/learning process. Instructional design is the analysis of training needs, the systematic design of teaching/learning activities, and the assessment of the teaching/learning process.

Planning instruction is an art with a scientific base. Each instructional planner will stamp his or her own personality and artful approach on the instruction, much the way an artist does. There are also basic skills of designing instruction that will ensure effective learning.

The flowchart in Figure 5.1 presents the three major components of the instructional design process and the supporting elements of each.

The flowchart is intended to serve as a guide in designing a course of study. It does not mean it is the only way to design courses. For instance, the instructor may find the first two or even

ANALYSIS	**DESIGN**	**EVALUATION**
• Assess Training needs • Conduct occupational analysis (Identify job competencies) • Determine learner characteristics • Establish levels of learning	• Write behavioral objective • Develop course outline • Develop lesson plan • Select instructional methods • Choose instructional materials • Develop testing tools • Allocate time	• Test for learner outcome • Evaluate instructional process

Figure 5.1 The instructional design process is composed of three major components.

four elements of the major component have been completed or that top management has defined those elements for the instructor. After thinking about what learning outcomes should occur, an instructor may reconsider and rewrite the objectives. In other words, these are the elements in the process of designing instruction, not necessarily the sequence. They are flexible elements and should be used with judgment. The remainder of the chapter will follow the flowchart elements.

ANALYSIS

Training Needs

Analysis or assessment of training needs is simply identifying the gap between what exists and what should exist. A needs assessment answers the questions: "Where are we today?" and "Where do we want to be in the future?"

A needs assessment generally reveals one or more problems:

1. If employees do not know, they have a lack of knowledge.

2. If employees cannot perform, they have a lack of skills.

3. If employees do not care, they have a lack of motivation or an attitude problem.

Each need can occur at different levels of the organization (department, company, or individual) (Figure 5.2). For example, the department purchases a new piece of equipment. All fire fighting personnel will require training in the use and maintenance of the equipment. Engine Company 4 is not working smoothly as a team. Mistakes have been minor so far, but teamwork skills and attitudes need improvement. In another example, an individual firefighter did not don the SCBA correctly, resulting in work time loss. Learning how to properly don the SCBA is essential for that firefighter. Needs are identified, unfortunately often through accidents, near misses, or injury. When the organizational level is determined, the instructor will know where to look for data, resources, and help to analyze needs or problems.

Gathering information to conduct a needs assessment can be done internally and externally. Internally, the instructor can review test scores and performance ratings; interview other instructors, supervisors, former students, or employees; obtain recommendations from management; examine company records; and merely observe what does and does not occur in and out of training. Externally, the instructor can interview people or distribute questionnaires at conferences or workshops. Another source is the National Fire Protection Association. It has conducted a national needs assessment and published the results in the professional qualifications. These are only examples of ways to collect needs assessment information. They should not limit the innovative instructor.

NEEDS OCCUR AT
- DEPARTMENT
- COMPANY
- INDIVIDUAL

Figure 5.2 Needs can occur at all different levels of an organization.

Finally, the instructor must judge whether the needs can best be met through classroom training, on-the-job training, closer supervision, development of policies or procedures, or redesign of a job. The instructor should realize that not all gaps, needs, or problems can be addressed by training, nor should they be.

When assessing training needs, considerations will be cost, practicality, administrative feasibility, and acceptability. Other people may need to be involved in decision making. Those who can help should be included.

The needs assessment step involves the following:

- Classifying what type of need is present (knowledge, skill, or attitude)
- Determining at what organizational level the need is occurring
- Judging whether training is the most appropriate means of closing the gap between what is and what should be.

If needs assessment identifies a problem that can be corrected by training, this is the motivation for continuing the development process.

Occupational Analysis

Occupational analysis is information about an occupation used to develop a detailed description of the qualifications necessary for a specific occupation, conditions for performance, and the duties and tasks involved. It is a dynamic document that changes as job demands change. Occupational analysis as used in other organizations may be referred to as job analysis or task analysis.

One of the purposes of occupational analysis is to determine what is to be taught. An occupational analysis is simply a listing of all tasks that must be done and the knowledge and skills required to adequately function in an occupation. These skills and knowledge are broken into specific tasks that personnel in the occupation are required to perform. An occupational analysis provides the instructor with a necessary framework of instruction tasks.

The occupational analysis is divided into divisions for better organization and easier use. The IFSTA validated **500 Competencies** is one example of an occupational analysis. Another example of the topic headings included in an occupational analysis is shown in Figure 5.3 on next page.

A *block* is a division of an occupation consisting of a group of related tasks with some one factor in common. In the fire service, a block could be firefighter levels I, II, or III, or it could be different positions such as driver/operator or fire officer. Blocks can also be organized according to fire department operations, such as equipment operation, water supply, fire control, and so on. Blocks are designated by Roman numerals.

CONTENTS

		Page
	Preface	i
BLOCK I	Organization	1
BLOCK II	General Maintenance	2
BLOCK III	Apparatus and Equipment Operation	4
BLOCK IV	Water Supply	7
BLOCK V	Fire Control	8
BLOCK VI	Salvage	10
BLOCK VII	Motor Vehicle and Pump Operation	11
BLOCK VIII	Emergency Care	13
BLOCK IX	Fire and Arson Investigation	14
BLOCK X	Public Relations	15
BLOCK XI	Fire Prevention	16
BLOCK XII	Fire Communication System	17
BLOCK XIII	Clerical Functions	19
BLOCK XIV	Related Mechanics	20
BLOCK XV	Fire Service Instruction	21
BLOCK XVI	Fire Service Supervision	22
BLOCK XVII	General Fire Administration	23
BLOCK XVIII	Fire Defense Administration	25
BLOCK XIX	Fire Personnel Administration	26

TOPIC HEADINGS FOR AN OCCUPATIONAL ANALYSIS

Figure 5.3 The occupational analysis is easier to use when it is divided into topics.

A *unit* is a breakdown of a block that brings the tasks into a more natural grouping. Units are designated by capital letters.

A *task* is a combination of jobs requiring the teaching and learning of psychomotor skills and technical information to meet occupational requirements. Tasks are designated by Arabic numerals. Each task may contain numerous jobs that must be taught and learned.

A *job* is an organized segment of instruction designed to develop psychomotor skills or technical knowledge.

THE OCCUPATIONAL ANALYSIS PROCESS

The first step in an occupational analysis is gathering information concerning the occupation. Information should include all of the following:

- A general description of the work to be done.

- The organizational setting and relationships in which the work is to be accomplished.

- The specific job to be carried out.

- The equipment, tools, and materials to be used.

- The working conditions and special hazards.

- The qualifications required to learn and perform tasks in terms of knowledge, skills, abilities, and personal characteristics.

- The expected manner or quality of performance.

- The process and expected outcomes of the performance.

Although all of the above information is important to conducting a thorough occupational analysis, the instructor should focus on the specific job to be carried out.

At this point, no job should be overlooked. Selection will come later. The first step in occupational analysis is to ask what, how, and why a worker does the job. The answers can be found from various sources: the NFPA standards, job descriptions, curriculum guides, and supervisors or employees.

The second step is to determine how often the job is performed, the relative importance of the job, and the complexity of the job. These three factors must be weighed with judgment. For example, a firefighter will sweep floors once a day, but may only rescue someone from a burning building once during a career. These factors are critical in selecting the necessary jobs to teach, because not all jobs will be used for instructional purposes.

The third step, selection of jobs for training purposes, should be based on the jobs performed by a large percentage of workers, performed frequently, critical to occupational accomplishment, essential to the performance of another job, or required for occupational entry.

The fourth step is to sequence the jobs according to some basic guidelines. Jobs should be developed from the simple to the complex. They should be arranged from the most frequently used jobs to the least frequently used jobs. Jobs should be sequenced from those needed most to perform other jobs to those less needed to perform later jobs.

The final step is to determine the related topics necessary to know in order to perform the operations within a given job.

As comprehensive as an occupational analysis is, it may still not meet the needs of every fire service agency. Specialized equipment, development of new methods, or the requirement of some special knowledge may necessitate the expansion of the occupational analysis. If necessary, the occupational analysis can easily be changed by simply adding tasks to any unit within the established format. If it is necessary to add total units, they should be added under the proper block and the instructor proceed from that point step by step.

The analysis process is the same whether it is used on the total occupation or is a part of the occupation. This process should be recognized as a powerful tool in work situations that can enhance anyone's career progress.

Learner Characteristics

Determining learner characteristics is the third element in the analysis component of instructional planning. This step includes identifying the intended audience. Although identifying the audience is often overlooked, it can help the instructor develop appropriate levels of instruction and meaningful teaching methods and aids.

If possible, the instructor should determine the general characteristics of the anticipated learners. Some of the characteristics of learners and typical resulting actions taken by an instructor will include the following:

Characteristic	Resulting Action
Academic/educational background	Adjust material to appropriate level of the student.
Personal/social characteristics	Use examples that speak to the lifestyle or culture of the student.
Learning ability	Anticipate a slow or gifted learner and adjust learning activities on an individualized basis.
Learning styles	Prepare lessons to reach as many senses as possible to accommodate as many styles as possible.
Previous experience or knowledge	Conduct a pre-test to assess where a student should begin new instructional content.
Student attitude	Assess reasons for student's presence in the class and provide appropriate motivation and incentives for learning.

Levels Of Learning

The fourth element in the analysis component of instructional planning is establishing levels of learning, also referred to as levels of instruction. "Levels of learning" refers to the depth of learning for a specific skill, attitude, and/or technical information that enables the student to meet the minimum requirements of the occupation. The purpose of classifying levels of learning is to determine the direction of the instruction and to assist in developing behavioral objectives.

The level of learning should be determined for each job listed in a course outline. In other words, levels of learning should be identified for each lesson plan developed. Lesson plans for teaching manipulative skills and technical subjects should only include one level of instruction.

When developing lesson plans and assigning levels of learning, the instructor should not attempt to teach all jobs at the highest level. This would be unrealistic and too time consuming. Instead, the instructor should identify the necessary level of learning by defining what a student needs to know to do a job. This is the reason the instructor must study the requirements of the job before assigning levels of instruction.

Levels of learning have been defined in many different ways. Three levels of learning are presented here; they are used to make planning and implementing instruction easier (Figure 5.4). This simplification is based on the extensive work of experts in the field of education who have supplied the foundational thinking that leads to these three levels.

LEVELS OF LEARNING
Level 1 ... Basic
Level 2 ... Intermediate
Level 3 ... Advanced

Figure 5.4 Levels of learning make planning and implementing class instruction easier.

LEVEL 1: BASIC KNOWLEDGE (BASIC) — This is the foundation level of learning. All other learning builds upon this level. At this stage, the student is acquiring new information (the cognitive domain), and determining the relationship of the information to his or her personal needs (the affective domain). Developing new skills (psychomotor domain) is not taught in the first level of learning.

At this stage, the instructor is an essential partner in the learning experience. The instructor channels new information to the student by teaching, making assignments, and guiding class discussion (the cognitive domain). The instructor must express the attitudes expected of the student, explain the reasons for learning the information and developing the skill, and model the proper attitude toward conducting the skill. The student must know what attitudes are expected on the job, understand why they are important, and see them exhibited consistently by the instructor to be able to integrate them (the affective domain).

As the student acquires information, the instructor confirms retention by having the student answer questions, take written quizzes, and participate in group exercises and discussions.

Evaluation of student progress is relatively easy at this stage, because the student is expected to do little more than memorize data. Objective tests, therefore, require the student to either recognize the correct answer or to supply (recall) the answer to a statement or question. Evaluation consists of observing the student actively listening, asking pertinent questions, making appropriate comments, and participating in group activities.

LEVEL 2: COMPETENT (INTERMEDIATE) — For the student to progress, foundation knowledge is connected to performance in the field. The student is required to apply previously learned information to prescribed problems. The student is then expected to make choices and to disregard irrelevant data. As the student applies knowledge to problem solving, the instructor switches emphasis. Instead of concentrating on the "how-to" mode used in Level 1, the instructor moves into a "why" mode, with explanations for actions taken. The "how-to" mode is not discarded, merely de-emphasized. Questioning focuses on confirming that the student understands concepts and their application to actual, day-to-day work. The student is asked to justify the approach taken to solve a problem. Therefore, the process becomes as important as the solution.

Skills are developed to an efficient level as wasted motion is reduced. Steps are consolidated into a smooth, fluid evolution, performed with little hesitation. Skills become second nature, so the student is thinking, not of the immediate step, but ahead to what follows.

At this level, attitudes should be forming and becoming a habit. Attitudes and values may not yet be firm or fixed. However, it will be more apparent that some thought is given to the expectations outlined in Level 1. Students will start voluntarily choosing to accept the attitudes and values as their own. The instructor should serve as a role model and exhibit the expected attitude since modeling is more effective than teaching for instilling attitudes.

In Level 2, testing becomes more complex because objective evaluation of understanding is more difficult. Written tests require the student to select and apply specific facts from a wide body of information, then apply them to similarly structured problems and situations. Some subjectivity is not only appropriate in this case, but often necessary to confirm understanding.

In subjects that require a shift in attitude from "noncommitted" to understanding and adherence to organizational and safety rules, evaluation becomes more difficult. Attitudes and values are generally observed during class activities and practical exercises. Verbal feedback is given during class. For example, if safety is not practiced by the student, the instructor might reinforce why it is important. In this case, observation and testing are designed to confirm that the student understands why the expected rules and attitudes toward them are important and how they can be demonstrated.

Manipulative tests at Level 2 focus on skill proficiency. For example, the student might be required to choose appropriate tools and use them to achieve an overall objective. To do this successfully, the student will need to consolidate several jobs (or skills) to complete an evolution. Evaluation of skill performance at this level focuses on the student's approach to the situation, ability to consolidate several skills into productive work, and speed with which the task is completed.

LEVEL 3: HIGHLY PROFICIENT (ADVANCED) — Bringing a student to the third level requires a significant investment of time by both the instructor and the student (Figure 5.5). This instructional period prepares the student to function in the field with little supervision. At this point, the student is expected to assume more responsibility for skill, attitude, and knowledge development. The instructor serves less as a teacher and more as a monitor and facilitator.

The instructor directs the student toward independent study of textbooks, periodicals, and other sources of information that will deepen conceptual understanding. The instructor challenges the student with increasingly complex and unique problems. The student is required to pull information from different subject areas, develop a solution based on accepted principles, apply it in an appropriate manner, and monitor and adjust the action as needed.

The instructor encourages student-initiated actions that are based upon the personal desires, feelings, attitudes, and values

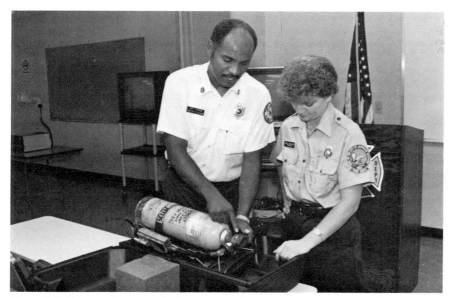

Figure 5.5 For a student to reach the advanced level, both the student and instructor must invest a considerable amount of time.

chosen by the student. At Level 3, the student's behavior will be controlled by those factors. Therefore, behavior can be a fairly good indication that the student has achieved the proper attitudes and values.

In Level 3, the instructor spends time observing student performance in emergency scene simulations, in evaluating the student's reasons for taking particular actions, and testing adaptability to quickly changing situations.

Evaluation at Level 3 is no longer strictly bound to the classroom or to the training ground. Because the student is expected to function independently with little or no supervision, performance must be observed in all types of situations, both simulated and real. This will confirm that the student is able to react, under pressure, to any emergency and perform competently and consistently. The evaluation system must, therefore, include written tests and exercises, single- and multi-company evolutions, and performance counseling and appraisals based on the day-to-day performance of duties.

DESIGN
Behavioral Objectives

The first step in the design component of the instructional design process is drafting clear, measurable objectives. A behavioral objective is a measurable statement of behavior required to demonstrate that learning has occurred. A behavioral objective will answer the question: "What is the student to know or be able to do as a result of learning or upon completion of the unit?"

Several other terms are used interchangeably with the term "behavioral objective." These include learning objective, instructional objective, learning or learner outcome, and performance objective. All terms refer to the learned behavior as a result of instruction.

A soundly constructed objective will provide a basis for designing instructional content, methods, and materials; for determining whether learning has occurred or instruction has been successful; and for organizing and motivating a student's own efforts toward learning. Therefore, objectives serve as a guide to learning, instruction, and evaluation. The instructor should:

- State the behavioral objective (Figure 5.6).
- Teach the behavioral objective (Figure 5.7).
- Test the behavioral objective (Figure 5.8).

DEVELOPING MEASURABLE OBJECTIVES

To be meaningful, objectives must be stated in terms of measurable performance. Further, they must be based upon the information gathered in the analysis component of the instruc-

Figure 5.6 Behavioral objectives should be stated clearly to the students so they will know what is expected of them.

Figure 5.7 The instructor must effectively teach the behavioral objective.

Figure 5.8 Testing the behavioral objective gives the instructor feedback on what the students have learned.

tional design process (Figure 5.9 on next page). Behavioral objectives must be specific, clear, stated at the beginning of each lesson plan, and given to the students, preferably in writing. When students are given the desired objectives, they know exactly what is expected of them to meet minimum acceptable standards, and they are motivated to meet these standards.

Behavioral objectives specify the conditions under which the behavior takes place, the expected behavior, and the standard of performance. An objective meeting all three of those criteria would indicate the following: conditions given (tools, information,

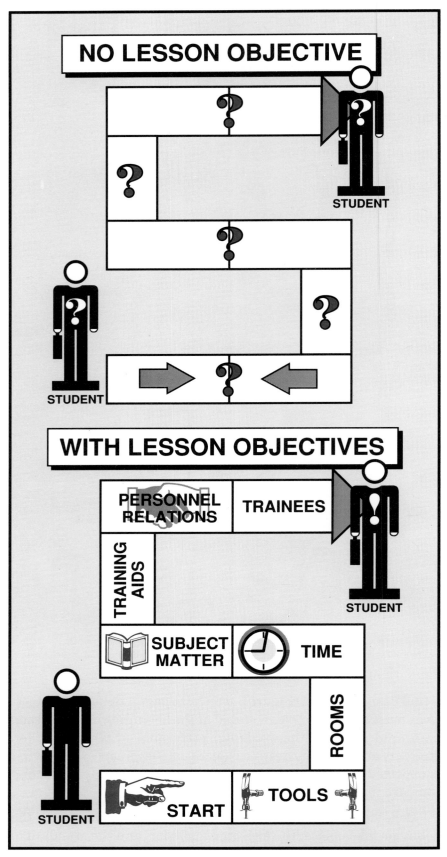

Figure 5.9 Objectives must be stated in terms of measurable performance to have meaning to the students.

or environment the student needs to perform a behavior), desired behavior (what the student should know or be able to do), and standard (how well the student must perform to show mastery of the desired behavior).

CONDITION—A properly prepared objective clearly states the given tools and equipment that the student is expected to use. This portion of the objective describes the important aspects of the work environment such as with what the student has to work, the tools to be used, whether notes or textbooks can be used, and physical conditions. For technical lesson plans, the conditions will be a written exam, outside assignment, or an in-class activity.

BEHAVIOR — An instructor cannot read the student's mind to verify the extent of understanding. It is only through some overt activity of the student that the extent of individual knowledge or skill can be determined. In the preparation of behavioral objectives the use of an action verb reduces ambiguity and aids the understanding of instructional intent. The behavioral objective must be specific and clear.

STANDARD — The performance standard indicates the level of performance a student must achieve to show competency or mastery of the desired behavior. The standard serves as a criterion for determining whether the student has achieved a satisfactory and safe level of performance, and describes the minimum acceptable level of performance.

Therefore, CONDITIONS describe with what the behavior is to be accomplished. BEHAVIOR describes what the student should know or be able to do. STANDARD describes how well the behavior is to be accomplished.

Using the C-B-S formula provides an easy, consistent way to write objectives. The first step in writing the behavioral objective is to list the key elements of an objective in descending order, leaving space for filling in the specifics:

CONDITIONS _____
_____
_____
_____

BEHAVIOR _____
_____
_____
_____

STANDARD _____
_____
_____
_____

First, let's look at the development of an objective for a psychomotor task. Suppose a job in a course outline is "How to make a 2 ½-inch nozzle connection using the over-the-hip method." The job title and level of learning indicate that the student "behavior" will be psychomotor or manipulative—to physically connect the nozzle to the hose.

A properly written objective for this psychomotor behavior would look like this:

> Given a length of 2 ½-inch hose and a 2 ½-inch nozzle (CONDITIONS), the student will connect the nozzle to the hose using the over-the-hip method (BEHAVIOR), completing all steps with 100 percent accuracy within 10 seconds (STANDARD).

Now let's develop an objective for a cognitive task. Remember: in these objectives, the condition, or given, is generally understood to be a written test, an in-class activity, or an outside assignment. Using as an example a job in the course outline that states: "students should know the purpose of pre-incident surveys," the objective could be written as follows:

> Given a list of purposes of pre-incident surveys (CONDITION), the student will demonstrate recognition of the purposes of pre-incident surveys (BEHAVIOR) by selecting 8 of the 9 listed purposes (STANDARD).

Writing measurable objectives for psychomotor tasks is simple because these tasks are all behaviors that can be readily observed. However, writing measurable objectives for cognitive skills is a bit trickier. How does the instructor observe that someone knows a principle or understands a concept when understanding and knowledge are not directly observable? For example, the instructor wants to know whether a person can add whole numbers, but the skill of adding whole numbers is not directly observable. For simple problems, whole numbers can be added mentally — that is, the student could add whole numbers without saying, writing, or displaying any outward action. Therefore, if the instructor wants to evaluate whether a student can add whole numbers, the student must be required to do something that can be observed. An example of this would be writing the answer to addition problems on a sheet of paper or saying the answers aloud. When writing measurable objectives, these activities are called "indicator actions."

AVOIDING COMMON PITFALLS

If the following guidelines are observed, several common pitfalls can be avoided when developing behavioral objectives.

1. Write an objective that refers to what the student will learn. Avoid stating what the instructor will do or what the in structional process will be. The purpose of an objective is to state what learners are expected to know or be able to do when they have completed a lesson. An objective that includes all three of the development elements also lets learners know what they are going to be given to accomplish the objective, how they are going to be evaluated, and what standard they must meet to show competency in that objective.

2. Write simple, straightforward objectives. Sometimes instructors become wordy or flowery, concentrating on flair and style at the expense of simple communication.

3. Be aware that the verb choice often reflects the learning level. Try to write objectives geared to the highest appropriate level, but also be aware that the learning process does not lend itself to a clear-cut hierarchy. Often, a fine line separates the various levels. See Table 5.1 (on next page) for examples.

4. Use clear, specific action verbs to describe the expected behavior. See Table 5.2 (on next page) for examples.

DEVELOPMENT OF OBJECTIVES

Instructors should not feel there is only one way to develop instruction. Individual instructors and departments should use the methods that work best for them. The National Fire Academy has still another method of developing objectives that uses the following elements:

A — Audience
B — Behavior
C — Condition
D — Degree

An objective is a statement of what a firefighter should be able to do after a training session. Objectives should be clear and easy to understand. Firefighters should know what they need to accomplish. Instructors, in turn, will know when firefighters have mastered the objective. A good objective will make all of the four elements — A,B,C, and D — clear.

An example objective would be:

"The firefighter will successfully don self-contained breathing apparatus in 60 seconds wearing full turnout gear according to department standard operating procedure (SOP)."

A — "The firefighter. . ."
B — ". . .will successfully don self-contained breathing apparatus. . ."

TABLE 5.1
EXAMPLES OF VERBS AND THEIR CORRESPONDING LEARNING LEVELS (Cognitive Domain)

LEARNING LEVELS	CORRESPONDING VERBS
Knowledge (LEVEL 1) (ability to recall or recognize information)	**Recall, List, Label, State, Recognize (communication or situation previously presented), Quote, Define, Name, Distinguish, Identify** ILLUSTRATION: The student knows that the formula for computing the area of a rectangle is A = L x W.
Comprehension (LEVEL 1) (ability to understand or use information within a limited context)	**Restate, Interpret, Explain, Illustrate, Discuss, Give an example of your own** ILLUSTRATION: The student can explain that in the formula, A stands for area, L stands for length, and W stands for width.
Application (LEVEL 2) (ability to use abstractions in particular, concrete situations)	**Apply a rule, guideline, or principle; Calculate, Measure, Predict trends, outcomes, or results; Solve problems** ILLUSTRATION: Given a rectangle 4 inches long and 3 inches wide, the student can determine that the area of the rectangle is 12 square inches.
Analysis (LEVEL 2) (ability to break information into its parts to clarify relationships)	**Test, Check, Inspect, Analyze, Determine conditions, properties, aspects; Verify, Divide, Classify** ILLUSTRATION: The student recognizes that if, for a given rectangle, either the length or width is increased, area will also increase.
Synthesis (LEVEL 3) (ability to create a new communication or concept through the examination of other communications or situations)	**Create, Design, Compare, Describe, Develop, (new information, not recital), Produce, Formulate, Devise** ILLUSTRATION: Knowing that the formula for the area of a rectangle is A = L x W, the student can develop a formula for computing the area of a parallelogram.
Evaluation (LEVEL 3) (ability to use standards and criteria to make judgments)	**Rank, Judge, Evaluate, Choose (based on standards), Diagnose, Appraise, Recommend, Decide, Justify** ILLUSTRATION: The student could choose from the following list which is the most accurate formula for determining the area of a right triangle. a. $(^1/2\ L) \times (^1/2\ W)$ b. $2 \times L \times W$ c. $\dfrac{L \times W}{2}$ d. $^1/4\ (L \times W)$

C — ". . .wearing full turnout gear. . ."

D — ". . .In 60 seconds according to department standard operating procedure (SOP)."

Firefighters working toward this objective know what they will need to accomplish to meet the objective. The instructor can tell whether or not firefighters have met the objective. Good objectives make it easier to accomplish training.

Course Description

A course description relates the basic goals and objectives of the course in a broad, general manner. It is designed to provide a framework and guide for the further development of the course, as well as a means of communicating the course content.

TABLE 5.2
PERFORMANCE TERMS FOR BEHAVIORAL OBJECTIVES

TERM	DESCRIPTION	EXAMPLE
Describe Synonyms: *discuss, define, tell how*	Reports the essential properties or characteristics of an object or event	**"Describe** the eight-step problem-solving process."
Define Synonyms: *describe, delineate*	Provides a description that gives the precise meaning or fundamental traits	**"Define** *scalar.*"
Identify Synonyms: *mark, match, choose, recognize*	Selects a named, described, or pictured item orally or by pointing to it, picking it up, labeling it, or marking it	**"Identify** ventilation tools in the illustration below."
Name Synonyms: *describe, delineate*	Supplies a title for objects, processes, events, principles, or people	**"Name** the type of heat that is generated by the splitting or combining of atoms."
List Synonyms: *write, arrange*	Recalls similar objects or events and records in a methodical or systematic arrangement	**"List** all of the hose fittings carried on the engine."
Order Synonyms: *arrange in order, list in order, sequence*	Arranges, rearranges, lists in sequence, or places in order	**"List in order** the three phases of fire."
Differentiate Synonyms: *distinguish, discriminate*	Recognizes as different and and separates into kinds, classes, or categories	**"Differentiate** among hazard classes for the following chemicals."
Classify Synonyms: *sort, arrange, group*	Puts into groups having common attributes, uses, characteristics, or functions	**"Classify** forcible entry tools according to their uses."
Construct Synonyms: *draw, make, build, design, create*	Makes an object, verbal statement, or a drawing	**"Construct** a floor plan for a pre-incident survey."
Apply Synonym: *use*	Uses a stated relationship or principle to perform a task or answer a problem	**"Apply** the progressive system to properly discipline an employee."
Demonstrate Synonyms: *show, perform, (any of various appropriate action verbs)*	Performs operations necessary to carry out a specified procedure	**"Demonstrate** the ability to properly set a roof ladder." or "Set a roof ladder."

The course description is based on the needs assessment, occupational analysis, levels of learning, and behavioral objectives. Once these items have been studied, the instructor can address the four elements of the course description:

1. Who is the intended audience? What rank, title, or position do they hold?

2. What is to be taught in the class? Or, what is the general course content?

3. What level of learning or degree of skill is to be attained in the course?

4. Where will the training be used most frequently (for example, at the fireground or fire station, in the public or internally)? How will the training be used most often (in a leadership role, as a subordinate, or with the public)?

The course description is a communication tool and sometimes serves as a marketing tool, depending on the purpose needed in a given situation.

Course Outline Development

Course outline development is the process of listing manipulative and technical jobs selected from an occupational analysis to meet predetermined teaching objectives (Figure 5.10). There are five major steps in developing a course outline:

- Determine the needs of the students
- Develop course objectives
- Identify jobs to be taught
- Organize jobs in a logical teaching sequence
- Establish tentative teaching times

Figure 5.10 The instructor should write down the overall plan of instruction.

This step in the design process may include research of several sources or may be an outline written from experience by the instructor.

Determine the needs of the students. This is the most important step in developing a course outline. In this phase, the instructor identifies what training the students will need to enter the occupation or to enhance their skills. The following are some ways to identify training needs:

- Determine the requirements of the student's position or rank or the requirements of tasks that must be performed.
- Identify the type of students who will be involved (for example, recruits, paid firefighters, or volunteers).
- Determine students' knowledge and skill levels by administering standardized tests or teacher-made diagnostic tests. These should be both written and performance tests.
- Observe students in the fire station, on the drillground, or on the fireground to determine their training needs.

Develop course objectives. Course objectives will determine the specific scope of the course. Course objectives are written by instructors for instructors. They serve as a guide for the instructor, they determine the specific jobs that must be taught, and they establish the basis from which student performance goals can be written.

Course objectives must describe what the students will do during the course. Although they are not necessarily written in measurable terms, they identify those behavior changes in the students that can be measured at the end of the course.

Identify jobs to be taught. The course outline must include a list of those jobs that the instructor will teach to help students develop needed skills. These jobs are correlated to the tasks in the occupational analysis. Some tasks may be performed by learning only one job, while other tasks cannot be performed until the student has learned many jobs. The number of jobs that must be included in the course outline depends on department and student needs. It also depends on the depth to which the instructor plans to develop student ability to perform tasks.

Organize jobs in a logical teaching sequence. Those jobs that a student must know first should be taught first. The teaching order sequence can be established either by the instruction-order method or by the production-order method. In the instruction-order method, instruction proceeds from the simple to the complex. In the production-order method, the teaching sequence is based upon the order in which jobs must be done under actual conditions and not necessarily in order of complexity.

Establish tentative teaching times. Although it is almost impossible to establish the teaching time for a course until lesson plans and other instructional materials have been developed, this factor should be taken into consideration early in course development. The amount of time available for a course of instruction will influence the course objectives, the teaching methods that will be used, and the lesson plans and instructional materials that will be required to conduct the course.

Lesson Plans

Since the purpose of instruction is to change the behavior of learners, the purpose of lesson plans is to guide the instructor through all the steps necessary for teaching a skill or knowledge in the proper sequence.

Four well-proven steps in teaching guide an instructor to do a complete job of instruction. These steps are preparation, presentation, application, and evaluation. These steps apply in any teaching situation, whether large or small (Figure 5.11).

FOUR-STEP METHOD OF INSTRUCTION

Step #1: Preparation — The first step to successful teaching is the preparation or motivation of the student. In this step the instructor must make a concentrated effort to reach the mind of the learner. Learning cannot take place until the learner is motivated. The instructor must get the students' attention. An effort must be made to arouse curiosity, develop interest, and create a desire to learn in the student.

This step should be used to build a teaching base. The student must associate every new idea or job to be learned with something already known. Therefore, the instructor should relate the lesson to the past experiences and knowledge of the students. In addition, the instructor should clearly state the objective of the lesson so students can understand what will be presented and what they are expected to learn. Although the motivation step is presented first when a lesson is taught, this step is prepared after the lesson is written when the instructor is planning instruction.

SAMPLE MOTIVATION/PREPARATION STEP TAKEN FROM A LESSON PLAN

Six firefighters died while fighting a fire in a one-story supermarket in Brooklyn recently. Not too long ago, two firefighters were killed in a Michigan restaurant fire when unvented products of combustion built up and exploded, causing rapid failure of the roof. More recently, in Los Angeles two firefighters died in separate fire incidents. Each was attempting to accomplish roof ventilation. An untold number of firefighters have nearly been trapped or have narrowly escaped when the roof of a building collapsed during a fire. Therefore, the points we are about to cover may save your life or that of your team member.

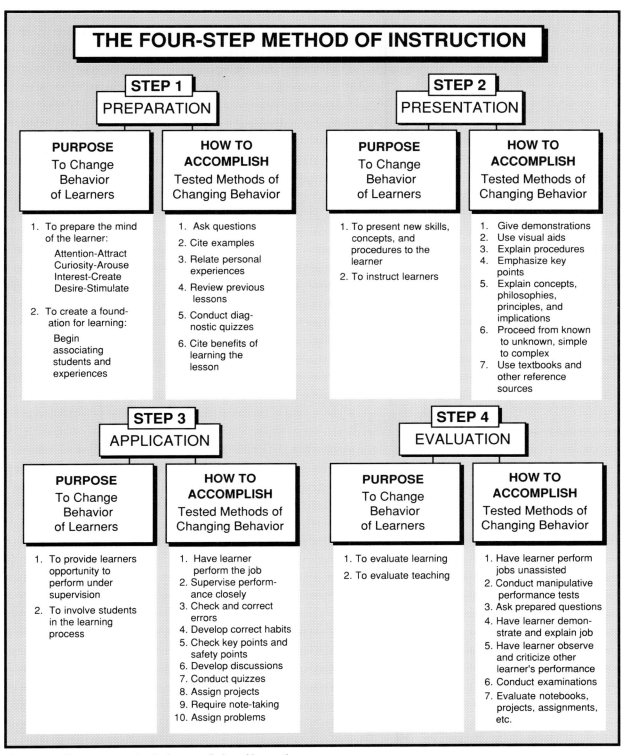

Figure 5.11 The course outline is the overall plan of instruction.

This lesson will deal with construction factors relating to ventilation. At the end of the lesson, you will be expected to identify the different types of roof construction and their specific hazards in relation to fire suppression activities. To do this, you will take a written quiz and must make at least 80 percent.

Step #2: Presentation — In the presentation step, new information, skills, and attitudes are presented to the student. This step involves explaining information, using supplemental training aids, and demonstrating methods and techniques. The presentation step requires resourcefulness and creativity on the instructor's part to capture and hold the students' attention. The material should be presented in a logical sequence by discussing one step at a time. Instructors should use the most effective combination of teaching methods and consider the appropriate use of teaching aids. Questions can be asked to get feedback and to re-emphasize important points.

Step #3: Application — The application step is a critical step in teaching. During this step the student has the opportunity to apply what has been learned. It permits the student to put into practice the new ideas, information, techniques, and skills that have been presented (Figure 5.12). It is important that students apply their new knowledge by performing the job or solving problems. The instructor may wish to have the student explain the key points as the job is performed. The instructor should supervise the performance closely, check key points and safety, and correct errors.

Figure 5.12 The application step allows the student to practice new techniques and skills.

Step #4: Evaluation — The purpose of the evaluation (testing) step is to see if the student has learned what was intended or whether the objectives of the lesson have been achieved. In the psychomotor domain, evaluation is the assessment of whether the student can do the job unaided or without supervision. During the application step the student performs the job under supervision; during the evaluation step the student works alone. This step gives the student an opportunity to demonstrate that the required degree of proficiency has been achieved. In the cognitive and affective domains, paper and pencil tests are typically used for testing purposes.

The four teaching steps have been discussed separately to point out the merits of each. The four steps should not be thought of separately, because they may be blended together (Figure 5.13). During the application step it may be necessary to repeat some of the demonstrations or explanations of the presentation. An experienced instructor will meet these needs without giving much thought to which step is involved. Considering the four steps separately when planning a lesson is convenient, however, and will prove beneficial while teaching. The four-step teaching procedure is not a rigid mechanical device that does not permit any deviation; it is an analytical outline around which a good lesson may be planned and taught.

Figure 5.13 The four teaching steps should not be thought of as separate from each other; they can be blended as needed.

In fact, one particular modification to this process is used extensively to teach toward deficiencies that a student or group of students may enter the class with. It is called prescriptive or remedial training. The prescriptive training system fully utilizes the four-step method of instruction, but in a different order:

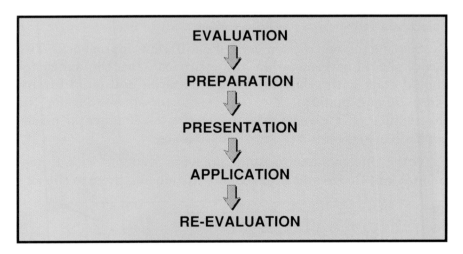

Historically, tests have been based upon what has been taught. In the prescriptive training system, learning activities are prescribed based upon skill deficiencies identified in a test at the beginning of a class. What, then, is the test based on? It is based upon carefully described performance standards (behavioral objectives) that, in turn, are based on the operations required to provide fire suppression services to the community.

Therefore, the prescriptive approach uses a different sequence in the four-step teaching method: evaluation, preparation, presentation, application, and re-evaluation. This allows the instructor to assess and prescribe the appropriate training to overcome learner deficiencies.

DEVELOPING THE LESSON PLAN

A lesson plan is a step-by-step guide for presenting a lesson (Figure 5.14). It serves as an outline of the material and procedures that the instructor intends to follow. The lesson plan ensures the efficient use of time and accurate coverage of the subject matter. It also contributes toward uniformity in teaching and avoids overlapping instruction.

Effective planning leads to effective teaching. Questions, learning difficulties, and student problems can be anticipated and solutions developed beforehand. Needed materials should be organized and ready for use before the instruction period. The lesson plan allows:

- Emphasizing specific information
- Effectively using training aids
- Staying within the required or allotted time
- Inclusion of proper and essential information

The essence of lesson planning is to develop instructional units that are most meaningful to the student and make effective use of time, space, and personnel (Figure 5.15). A lesson may vary in length from a few minutes to several hours, depending upon the objectives set.

Sequencing instruction can take different approaches. Two basics should be included in any approach. One, the instructor should start with relatively simple material and build toward more difficult material. Two, the instructor should develop the instruction so it can be adapted to the developmental needs, capacities, and maturity levels of students.

Often, instructional content presents students with fundamental knowledge and skills before introducing them to the jobs

Figure 5.14 A lesson plan provides a guideline for presenting a lesson.

Figure 5.15 There are many things to consider when developing a lesson plan.

in which they will be using these skills and knowledge. Because this type of sequencing is out of context of the related jobs, the fundamentals are less meaningful and harder to learn. To reduce student failure rates, "functional context sequencing" is recommended. This sequencing approach is characterized as follows:

- The student is given an overview of the entire job. The purpose of the job is explained to give the student an understanding of why it is being conducted.

- Instructional topics are organized so the relevance of each step to the whole job can be immediately demonstrated to the student.

- A whole-to-part sequence is followed in teaching the functions of equipment.

- Each student learns a series of jobs in steps. Each new job requires the student to master new knowledge and skills.

- The reverse process, the parts-to-whole sequence, can be introduced to enhance comprehension by the student.

Lesson Plan Format

Because fire service instructors must develop and teach lessons that address all three learning domains—affective, cognitive, and psychomotor—there are many different lesson plan formats. A typical format requires the instructor to break the lesson into two broad components: technical lessons and manipulative lessons. The technical lesson plan, or information presentation, deals with theory and technical knowledge—such as fire chemistry and behavior—that are essential to skill development. This aspect of the lesson relates primarily to the cognitive domain of learning. The manipulative lesson plan, or practical demonstration, deals with psychomotor skills such as climbing ladders or extinguishing fires. The affective domain of learning evolves naturally from either component.

Format For A Technical Lesson Plan (Information Presentation)

While there are many different ways to format the technical lesson plan, certain basics, as shown in Figure 5.16, are common to all. These basics specifically call for identifying all objectives (both cognitive and psychomotor) that the student must achieve and for providing a framework within which those objectives can be taught most effectively. If this format is followed, it is impossible to fall into the trap of developing a lesson based on what the instructor wants to do, rather than what students need to learn. A description of each part of the information component of the lesson plan follows:

- TOPIC — A short, descriptive title of the information to be covered. The title should somewhat limit the content of the lesson, but not be so brief that it does not describe the lesson content. Topic titles are taken directly from the course outline.

- TIME FRAME — The estimated time it will take to teach the lesson. Sometimes time frames are established for each objective so that the instructor has better control of pace.

- LEVEL OF INSTRUCTION—Based on job requirements, the learning level (1, 2, or 3) is established and given here.

- BEHAVIORAL OBJECTIVE(S) — A description of the minimum acceptable behaviors that a student must perform by the end of an instructional period.

- MATERIALS NEEDED—A list of all materials in sufficient quantity needed to teach the number of students involved. This section of the lesson includes any preparation, planning, or activities that the instructor needs to complete before delivering the lesson.

- REFERENCES — Specific references, including page numbers, that the instructor must study to teach the lesson. When lesson plans are developed by curriculum specialists, this section may also include a list of references consulted in the development process.

- PREPARATION — Step 1 of the four-step method of instruction. Here the instructor establishes relevancy by introduing the topic and objectives of the lesson. The material and methods outlined here are designed to motivate the student to learn the information that is to be presented.

- PRESENTATION — Step 2 of the four-step method of instruction. The instructor lists in order the information to be covered, how it will be covered, and what the instructor must do to teach the lesson.

- APPLICATION — Step 3 of the four-step method of instruction. The instructor supplies activities, exercises, and jobs so that the student can apply the information taught in the lesson. The application step does not always have to follow the presentation step. Often the instructor integrates application steps into the presentation, allowing students more immediate involvement in the learning process.

- LESSON SUMMARY — Restatement of or re-emphasis on important information presented in the lesson. Summaries serve to clarify uncertainties, prevent misconceptions, and increase learning and retention.

- EVALUATION — Step 4 in the four-step method of instruction. The instructor develops and/or administers a test to assess the amount of cognitive and affective learning that took place and to determine whether the behavioral objectives were achieved.

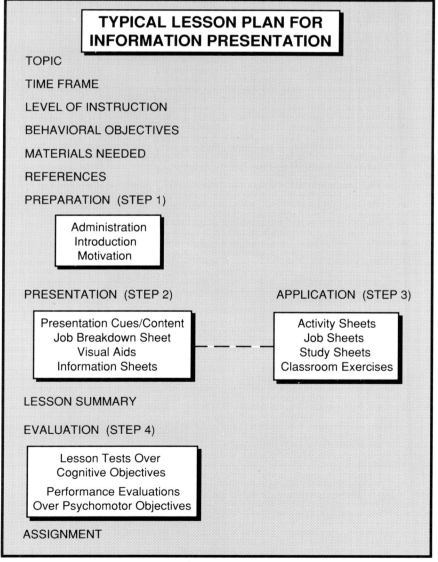

Figure 5.16 A typical lesson plan for informationl presentation.

- ASSIGNMENT — Additional work that must be performed by students outside of class to reach the established behavioral objectives or to prepare for the next lesson. There need not always be an assignment.

Format For A Manipulative Lesson Plan (Practical Demonstration)

While the technical lesson is designed to teach cognitive skills, the manipulative, or practical, lesson is designed to teach psychomotor skills. Because of this, the format of this lesson component differs considerably from that for the information component. However, the basic four-step method is common to both. Elements used in the development of the manipulative lesson plan include the following:

- JOB OR TOPIC — A short, descriptive title of the information to be covered. The title should somewhat limit the content of the lesson, but not be so brief that it does not describe the lesson content. Job or topic titles aretaken directly from the course outline.

- TIME FRAME — The estimated time it will take to teach the lesson.

- LEVEL OF INSTRUCTION — Based on job require ments, the learning level (1, 2, or 3) is established and given here.

- BEHAVIORAL OBJECTIVE(S) — A description of the minimum acceptable behaviors that a student must perform by the end of an instructional period.

- MATERIALS NEEDED — A list of all materials and equipment needed by the instructor to teach the lesson.

- PREPARATION — Step 1 of the four-step method of instruction. Here the instructor establishes relevancy by introducing the topic and objectives of the lesson. The material and methods outlined here are designed to motivate the student to learn the information that is to be presented.

- PRESENTATION — Step 2 in the four-step method. The job breakdown, listing the operations and key points for the job, is inserted and taught here.

- APPLICATION — Step 3 of the four-step method, where the student practices performing the operations in the job breakdown sheet under supervision.

- EVALUATION — Step 4 in the four-step method, in which students demonstrate how much they have learned. They must perform the job without assistance from the instructor.

- ASSIGNMENT — There need not always be an assignment. However, learners forget new skills and sequences quickly unless they are given the opportunity to practice them. For example, an assignment such as knot tying may be given to allow the student to practice knot tying procedures necessary to perform a job.

JOB BREAKDOWN SHEETS

The job breakdown sheet is used to break a job down into parts by listing the operations and key points (Figure 5.17). A job sheet is developed for each of the jobs (psychomotor skills) specified in the course outline. The purpose of these sheets is to provide the instructor/student with the sequences and details necessary to teach/learn a job that includes both motor skills and required

---

JOB: HOW TO COUPLE HOSE (STRADDLE METHOD)

Name: _____ Date: _____

OPERATIONS	KEY POINTS
1. Straddle hose	1a. 12" behind female coupling b. Facing coupling
2. Pick up female coupling	2a. With working hand b. Palm down c. Fingers inside
3. Pick up male coupling	3a. With Free hand b. Palm up c. On coupling and hose
4. Inspect coupling	4a. Start while picking up b. Inspect gasket c. Inspect threads d. Inspect shape e. Check for obstructions f. Inspect visually or by feel
5. Clamp knees	5a. On hose 12" behind female coupling b. To secure hose and coupling
6. Set threads	6a. Coupling together and level b. Counterclockwise optional c. Higbee cut
7. Connect coupling	7a. With working hand b. Counterclockwise optional c. Hand tight
8. Lay connection down	8a. Gently b. With hose straight

Figure 5.17 A typical job breakdown sheet.

knowledge. The required knowledge, or key points, is often referred to as "knowing units," because failure to know this material prevents successful completion of the job. Key points are listed on the right-hand side of the job sheet. Motor skills are often referred to as "doing units." These units are listed on the left-hand side of the job sheet.

An *operation* is a step or the smallest aspect in performing a job. Operations are ordered with Arabic numerals.

A *key point* is a factor that conditions or influences the performance of an operation. It may include the knowledge or understanding needed to perform the operation correctly.

Developing A Job Breakdown Sheet

Develop a job breakdown sheet in the following manner:

Step 1: List the job to be done (this becomes the title).

Step 2: Divide the page into two columns.

Step 3: Head the left column "Operations" or "Doing Units" (the actual motor skills).

Step 4: Head the right column "Key Points" or "Knowing Units" (those pieces of knowledge without which the operations cannot be safely or accurately performed).

Step 5: Under "Operations," list in sequence the steps of the job, using action verbs (such as grasp, push, turn, lift, and so on).

Step 6: Under key points, list cautions, warnings, safety factors, and conditions essential for performing the job operations.

A completed job breakdown sheet lists the step-by-step procedures for doing a job, in sequence, and the key points that the instructor must stress while teaching the job. The content and format of a well-prepared job breakdown are shown in the sample worksheet that follows (Figure 5.18).

Time

To adequately plan instruction, it is helpful to estimate the time it takes for each lesson or unit and to total these estimations to determine approximately how much time is needed to conduct the course. The lesson plan should be thought of as a guide showing the way students and instructor will spend their class time, rather than as a document that precisely dictates what must happen during each instructional minute. Because it is impossible to make exact predictions about the time, the plan should allow for flexibility. Once the instructor conducts the course, instruction can be added or deleted and changes made as needed.

PERFORMANCE EVALUATION

JOB SHEET 1-1
COUPLE HOSE USING THE STRADDLE METHOD

STUDENT'S NAME _____ DATE _____

EVALUATOR'S NAME _____ ATTEMPT NO. ___

INSTRUCTOR'S NOTE: To show competency in this job, the student should be able to perform each operation of this procedure in no more than two attempts. If the student is unable to achieve this competency, assign review procedures and schedule another evaluation date.

OPERATION	1st Try	2nd Try	N/A	Comments
1. Straddled hose 12" behind female coupling				
— faced coupling				
2. Picked up female coupling				
— with working hand				
— palm down				
— fingers inside				
3. Picked up male coupling				
— with free hand				
— palm up				
— on coupling and hose				
4. Inspected coupling visually or by feel				
— began while picking coupling up				
— inspected gasket				
— inspected threads				
— inspected shape				
— checked for obstructions				
5. Clamped knees on hose 12" behind female coupling				
6. Set threads				
— held coupling together and level				
— located Higbee cut				
7. Connected coupling hand tight				
8. Laid connection down with hose straight				

Figure 5.18 Students can use job breakdown sheets to prepare for a performance evaluation.

APPLICATION TOOLS, INSTRUCTIONAL AIDS, AND TESTING INSTRUMENTS

Once the elements of the lesson plan are outlined and completed, the instructor's next steps are to:

- Develop application tools to enable the student to apply the lesson content.

- Create instructional aids to clarify and reinforce learning.
- Design tests to assess the learning that has occurred.

    NOTE: Each of these aspects is addressed in greater detail in separate chapters: Chapter 6, Presenting the Instruction; Chapter 7, Training Aids, and Chapter 8, Evaluation and Testing. The remainder of this chapter will cover the most common types of application tools, namely, information sheets, student worksheets/activity sheets, and study sheets.

Information Sheets

An information sheet — sometimes called a "fact sheet" — is a handout that contains supplementary material related to the course text. Information sheets have the same purpose as study sheets. However, the study sheet tells the student where to find needed information, whereas the information sheet provides the student with that information, explained and illustrated in detail. A good information sheet encourages the student to learn, provides the necessary information, and tests the students' understanding of the material (Figure 5.19). Information sheets are usually developed for one of the following reasons:

- The information is unavailable to some of the students because the number of textbooks is limited.
- To get the information the student would have to consult a number of texts and it would be difficult to obtain all of these texts.
- The information is not available in any prepared text.

To develop an information sheet, take the following steps:

Step 1: Create a title that reflects the subject area to be studied. Key the title to the lesson.

Step 2: Write a brief introduction that explains the importance of the information and encourages the student to read and study the information presented.

Step 3: Present the information in outline form, charts, tables, or illustrations. Type the copy so that it is easy to read and follow. Illustrations can be inserted in the text or placed on a separate page.

Step 4: Develop test questions — either separately or as part of the lesson test — to assess whether the lesson objective was achieved (Figure 5.20 on next page). The test questions can be used to stress the important points in the information presented. They also permit the instructor to check on the student's comprehension. Use thought-provoking questions and include a sufficient number to cover the information thoroughly.

Student Worksheets/Activity Sheets

Student worksheets, or activity sheets, grow out of the lesson's information presentation (Figure 5.21 on next page). They may either support or be listed as the lesson's objectives. Those that support the lesson's objectives are said to be appropriate practice.

The main purpose of student worksheets is to give students a chance to use higher order thinking skills in both the cognitive and affective domains.

INFORMATION SHEET

TOPIC: Nozzle Flows and Equivalent Tip Size Information Sheet.

INTRODUCTION: The amount of water a pumper is flowing is an important part of the pump operator's job. Through knowledge of the various tips and nozzle pressures used in our department, the pump operator is better able to determine flow and use the pumper to its capacity. There has been some confusion among department personnel as to the rated output of the various size nozzle tips used in this department. Also, which size open tips our fog nozzles and fog tips would be compared to. The following information shows all nozzle tip sizes used by this department, their rated output, and what size open tips compare with the fog nozzles and tips.

DISCHARGE FROM OPEN TIPS

The following tips are open tips used on the 2 1/2" nozzles and master stream appliances.

Tip Size	GPM at 50 PSI	GPM at 100 PSI
2"	841	1189
1 3/4"	643	909
1 1/2"	472	667
1 3/8"	396	560
1 1/4"	326	461
1 1/8"	265	374
1"	209	295

FOG NOZZLE EQUIVALENT

Description	Rated GPM at 100 PSI	Equivalent to (open tip)
Master Stream Fog Nozzles	500	1 1/4"
2 1/2" Revolving Fog Nozzle	400	1 1/8"
2 1/2" Fog Nozzle	300	1"
Bressnan Type Distributor	280	1"
1" Fog Nozzle	28	5/16"
3/4" Fog Nozzle	18	1/4"
1 1/2" Fog Nozzle	75	1/2"

Figure 5.19 A typical information sheet.

INFORMATION SHEET TEST

NAME _____

DATE _____

Directions: Write the answer to each question in the
space provided for the questions.

1.   How does the pump operator determine the amount of
water the pumper is flowing?

2.   List the tips and their flow that we use on our $2^1/_2$"
handlines.

3.   Approximately what open tip size would a Bressnan
distributor be equivalent to?

4.   Could a 1,000 gpm pumper pump to $1^3/_4$" tips at 50 psi
nozzle pressure? Why?

5.   Why does the pump operator need to know the
equivalent tip sizes?

Figure 5 .20 Example of an information sheet test.

Worksheets/activity sheets allow students to apply rules,
analyze situations, evaluate objects or situations, or use several
skills together (such as writing a report on safety equipment
trends). Those that deal with the affective domain may not
directly support any one lesson objective but may instead support
several objectives. These activity sheets may deal with students
beliefs, attitudes, or appreciation of the subject area. Use the
following steps to develop an activity sheet:

Step 1:   Create a title that reflects the subject area to be studied.
Key the title to the lesson.

WORKSHEET/ACTIVITY SHEET 1-1:
EVALUATE YOUR LISTENING ABILITY

MATERIALS NEEDED:
1. Partner
2. Pencil and paper

INTRODUCTION: How well do you listen? Or do you only hear? One of the most basic ways a supervisor shows a worker he or she cares is by listening to the worker and trying to understand what he or she says. Listening and understanding go hand in hand. If you listen without understanding, then you are hearing but not really listening. As a company officer, it is critical that you learn to listen effectively. And, yes, listening **can** be learned. Before learning listening techniques, use this activity sheet to evaluate your present listening ability. Find out if you are a listener or hearer.

DIRECTIONS: Evaluate your listening ability by applying the guidelines below to the situations described on the following page. You will need a "feedback" partner to help you evaluate your listening skills. Your partner will rate your responses to the situations on a scale of 1 to 10, with 10 being the highest rating. After you have been rated on each of the listening situations, add your ratings. A score of 90 to 100 indicates EXCELLENT listening skills; a score of 80 to 89 is GOOD; 70 to 79 is FAIR; and 60 to 69 is POOR. If you score 79 or below, complete Activity Sheet 1-2.

---

1. Give your undivided attention to what the worker is saying.

2. Face the person you are listening to and look at him or her attentively.

3. Let the person know you are following the conversation by nodding occasionally or saying "I see."

4. If there are moments of silence, wait patiently to let the worker collect his or her thoughts or think over what has been said.

5. Think of what the other person is saying, not of what you will reply.

6. Do not interrupt when something the other person says triggers an idea. Wait until the person is through speaking.

7. Listen not only to the content of what the person is saying, but also to the **feelings** the person seems to have about the subject.

Figure 5.21 A typical student worksheet/activity sheet.

Step 2: List all the materials needed, including resources that the student will need to complete the assignment. List titles and specific page numbers of books, magazines, or other reference material you want the student to study. Provide complete information so the student will have no difficulty locating these resources.

Step 3: Write a short introduction that will encourage the student to complete the assignment satisfactorily. Include the following: 1) discuss the skill in general, telling how the activity relates to the subject or training area and lesson objective, 2) explain why learning the skill is important to the student, and 3) tell how the worksheet/ activity sheet will help the student master the required skill.

Step 4: Provide directions, telling students how to complete the activity sheet assignment.

Step 5: On a separate sheet, include answers (or suggested answers), if applicable.

Study Sheets

A study sheet is a written instruction sheet that arouses student interest in a topic and explains exactly what must be studied. Study sheets may be used as part of group instruction or for self study. A self-test may be included to allow the student to measure how well he or she understands the material. Use the following steps to develop a study sheet:

Step 1: Create a title or state the topic that reflects the subject area to be studied. Key the title to the lesson (Figure 5.22).

Step 2: List all the materials needed, including resources that the student will need to complete the assignment. List titles and specific page numbers of books, magazines, or other reference material you want the student to study. Provide complete information so the student will have no difficulty locating these resources.

Step 3: Write a short introduction that will encourage the student to complete the assignment satisfactorily.

Step 4: Present the study information in a format that helps the student learn the material.

Step 5: Provide complete information so the student will have no difficulty locating the study materials.

Step 6: Use a separate sheet of paper for the study sheet test if one is to be included. Test questions included as part of the study sheet should assess the student's understanding of all aspects of the subject and require the student to think. There should be enough questions to cover the study sheet thoroughly.

Evaluation

Evaluation is the systematic collection of information for judging the value of something. The purpose of evaluation is to improve the process or product of any endeavor. It is the way the instructor determines whether the lesson's behavioral objectives

---

STUDY SHEET

TOPIC: Pumper Tests

MATERIALS NEEDED:
1. Fire Company Apparatus and Procedures. L. Erven, 3rd edition, Beverly Hills, California, Glencoe Press,1979.
2. Pencil and paper.

INTRODUCTION:
The testing of pumpers allows the capabilities of the pumper to be seen by the pumper operator. This information is also used to determine the acceptance of new pumpers and to compare a pumper's performance from year-to-year. This study assignment will show you the type tests performed and help you learn to perform these tests.

ASSIGNMENT:
1. Study pages 151-162, Fire Company Apparatus and Procedures.
2. Answer all questions on the following page and turn in on _____.
   (date)
3. This material will be covered in a written examination on _____.
   (date)

Figure 5.22 Example of a study sheet.

have been achieved. Testing evaluates both the process and product of the objective. Take, for example, a psychomotor objective that reads, "Demonstrate the ability to tie a half hitch." The instructor observes and evaluates the process—the steps taken in the correct sequence—and the product—the completed knot. Most cognitive objectives are tested only for the product or result of a thought process because the way in which a person arrives at a conclusion cannot be observed.

Evaluation is covered extensively in Chapter 8, but must be considered early and applied to each step of the instructional development process. This is particularly true when shaping the lesson's behavioral objectives. When developing the test instrument(s), the instructor can assess the completed lesson plans to determine whether they meet the principles of learning. Often, adjustments and revisions are made at this time.

Informal classroom evaluation can be the simple attention to whether the lesson plan worked in the classroom. The instructor can sense whether the students are grasping the content and skills. The instructor will also feel comfortable or uncomfortable with the presentation methods and results of instruction. These impressions will come with experience. Never underestimate gut-level reactions, which may lead to improved instruction.

SUMMARY

Planning instruction is a vital task in any teaching/learning process. Each instructor approaches the instructional development process differently, but there are basic skills that all can learn to employ. The three major components of the instructional design process are analysis, design, and evaluation. Analysis consists of identifying specific training needs, job characteristics, and student characteristics. Design consists of writing behavioral objectives, developing the necessary course materials, and selecting instructional materials. The final step in the instructional design process is evaluation. Evaluation consists of testing learner outcomes and evaluating the instructional process.

Each step in the instructional development process is important. For instance, the instructor should not attempt to design instructional materials without first doing an analysis. Taking time to make a thorough analysis makes designing easier and more effective. Also, evaluation should uncover any problems in the teaching/learning process, and the instructor should not hesitate to make changes if they are needed. The seasoned instructor will always be alert to ways to improve or update instruction, or may even eliminate some instructional information or technique that is no longer effective. Most of all, the instructor should be flexible in today's changing world of instructional needs, content, methods, and techniques.

SUPPLEMENTAL READINGS

Finch, Curtis R., and John R. Crunkilton. *Curriculum Development in Vocational and Technical Education*: Planning, Content, and Implementation. Boston: Allyn and Bacon, Inc., 1979.

Fryklund, Verne C. *Occupational Analysis: Techniques and Procedures*. New York: Bruce, A division of Benziger Bruce and Glenco, Inc., 1970.

Kemp, Jerrold E. *The Instructional Design Process*. New York: Harper & Row Publishers, 1985.

Mager, Robert F. *Preparing Instructional Objectives*. Revised 2nd ed. Belmont, California: Pitman Learning, Inc., 1984.

Mager, Robert F., and Kenneth M. Beach. *Developing Vocational Instruction*. Belmont, California: Fearson Publishers, 1967.

Presenting
the Instruction

This chapter provides information that addresses performance objectives in NFPA 1401, *Fire Service Instructor Professional Qualifications* (1987), particularly those referenced in the following sections:

NFPA 1401

Fire Service Instructor

3-1.1 (e)(f)(k)

3-1.2 (l)(q)(r)

3-4.1 (e)

3-7.1

3-7.2 (a)(b)(c)

3-9.1

3-9.3

3-10.4 (b)(c)

4-6 (b)

4-8

Chapter 6
Presenting the Instruction

The manner in which the instructor presents information to the students will have an effect on how the students learn. There are many different ways information can be presented. The effective instructor will be able to determine which method or methods are most appropriate for each topic. Presenting information is much more than just getting up and lecturing to the students. Instructors should not be afraid to be creative and innovative when presenting information to students.

Management of the instructional activities is the responsibility of the instructor. Conditions must be established that will give the students the best possible opportunity to learn. The instructor should keep the students informed about the activities of the class by providing a session guide or calendar of events. Another important consideration for successful instruction is providing a good learning environment. This requires planning, sound organization, and continuous management (Figure 6.1).

Figure 6.1 Management of instructional activities requires planning and organization.

POLICIES

Clearly defined policies make class management easier. An instructor should establish policies and inform the students of them before beginning the first lesson. These policies should be in writing. Students feel more secure when they know what is expected of them. With clearly defined policies, certain discipline problems may be avoided from the beginning. The instructor should set policies on absenteeism, tardiness, class participation, evaluation, and assignment due dates.

If attendance taking is required, the instructor should not waste time calling each student's name every class session. Associating names with faces makes it easy to take attendance at a glance. Nameplates are helpful for the instructor. If this method is not possible, the instructor can pass a sheet around and have each student sign it. After this practice is established, it becomes routine and does not create a distraction.

CLASS DISCIPLINE

Once the policies of the class have been established, the instructor should make sure that students adhere to them. An instructor who ignores small infractions may have severe discipline problems later. If a student does break a rule, discuss the situation as soon as possible. Always try to discipline in private because embarrassing a student in front of peers can cause resentment. Be tactful but firm when discipline is needed. Generally, a tactful handling of the situation will produce the desired results. If a threat of punishment must be made, follow through immediately; otherwise, the instructor's authority is weakened and may be challenged in the future.

It is critical to document any level of discipline administered by the instructor. The date, time, circumstances, person or persons involved, and action taken by the instructor are minimal pieces of information needed to adequately document any disciplinary actions. Documentation is important because if progressive discipline is needed in the future, the instructor can verify a background of leading factors and previous actions.

SESSION PLANNING

The basis for session planning is the course outline. It lists in sequence the topics and jobs that will be taught and tentative teaching times. This guideline enables the instructor to schedule or plan the activities that must occur in each class session.

The correlation between classroom and field activities must be considered during session planning. The instructor must determine whether it would be best to teach a job in the classroom or on the drillground. If a particular job requires both classroom instruction and outside activity, thought must be given to maintaining the learning process during moves between each instruction area. Deciding when such moves should occur should also be considered. The instructor must also schedule time for reviews, tests, individual assistance, breaks, and for explaining assignments. In addition, scheduling for special activities, such as guest speakers, field trips, films and other visual aids, or extensive training exercises (tower work, fire fighting, or heavy rescue work), must be included in session planning.

Guidelines for organizing class sessions can be prepared by developing a session guide or a calendar of events. This calender is a list of the activities, topics, and jobs that will be covered in each class session, and what will occur on each date the class meets (Figure 6.2). It usually lists assignments and their due dates, any special or unusual activities, such as guest speakers or field trips, and other pertinent information.

CALENDAR OF EVENTS
INSTRUCTOR TRAINING

DATE	SESSION	SUBJECT TO BE COVERED
1-3	1	1. Course Introduction 2. Instructor's Role 3. Instructor's Responsibilities 4. Equal Opportunity and Affirmative Action Programs 5. Characteristics of Good Instructors
1-10	2	1. Instructional Terms 2. Student Motivation 3. Learning Influences 4. Occupational Analysis
1-17	3	1. Learning Process 2. Individual and Group Behavior 3. Occupational Analysis 4. Levels of Instruction
1-24	4	1. Behavioral Objectives 2. Job Breakdown Sheets 3. Lesson Plan Development 4. Four-Step Method of Instruction
1-31	5	1. Organizing the Learning Environment 2. Principles of Instruction 3. Methods of Instruction
2-7	6	1. Types and Uses of Training Aids 2. Utilizing Training Aids 3. Training Aid Development

Figure 6.2 A calendar of events lists what will occur on each date the class meets.

MAINTAINING CONTINUITY OF INSTRUCTION

Despite lesson planning and classroom or drillground preparation, situations will occur that can disrupt class presentations. Planning class sessions for fire service training activities must include planning for unusual occurrences. Classes may be interrupted by alarms, special details, unexpected maintenance needs, or sudden changes of priorities. Alternative plans should be developed and ready for use. Unanticipated situations may develop, however, including:

- Emergency calls
- Temporary drops in attendance
- Malfunction of equipment
- Failure of guest speakers to appear
- Lack of student interest in a particular subject or activity
- Failure of students to perform as planned
- Inclement weather

An instructor who gives adequate attention to the variables involved in class presentation will be prepared to adjust to unusual conditions. The instructor who has alternative plans will be able to immediately shift to another activity.

PHYSICAL SETTING

The physical setting of the classroom is a factor that should receive attention. Valuable class time is lost if the instructor has to move chairs, readjust the air conditioning, or otherwise alter the environment. All these preparations should be completed before class (Figure 6.3).

Figure 6.3 Instructors should be prepared for each class session before the class begins.

Many fire service instructors are forced to conduct classes in rooms and areas that leave much to be desired. Through planning, the instructor can usually eliminate or at least minimize some of the problems created by poor classroom facilities. The instructor should make sure that suitable chairs and tables or desks are available so students can easily take notes. All too often, instructors overlook this important aspect of learning when they have to conduct class sessions in recreation rooms, apparatus rooms, or dormitories.

Make certain the classroom temperature is comfortable and that lighting and ventilation are adequate. The seating arrangement should suit the type of instruction planned (Figure 6.4). If visual aids are to be shown, arrange the chairs so each person can see the screen. A good arrangement for a small, participative group is to place the chairs in a circle or semicircle.

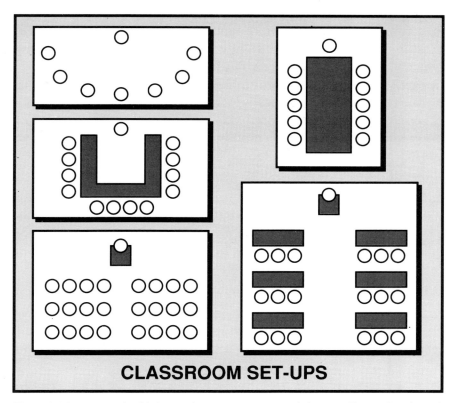

CLASSROOM SET-UPS

Figure 6.4 Instructors should plan seating arrangements to suit the type of instruction planned.

If any audiovisual equipment is to be used, make sure it works and is ready. Extra projection bulbs and take-up reels should be on hand along with the necessary extension cords and adapters. Arrange handout material and other training aids in the order in which they will be used. To avoid distraction, keep them covered or out of the way until they are ready to be used.

Fire service work requires much outdoor training. Such training may be conducted at drill towers, in fire station yards, on streets, in parking lots, and at target hazards. Developing and maintaining a suitable learning environment under these conditions is not easy. Some factors that limit teaching and learning efficiency in outdoor settings are:

- It is more difficult to control a group.
- There are more distractions.
- Notes are difficult to take and handle.
- Students may find it difficult to see and hear.
- Weather conditions may affect the teaching-learning environment.

Because outside instruction involves unique environmental considerations the instructor should plan, organize, and manage field training activities even more carefully than classroom sessions.

In the classroom, control is usually maintained as a result of confinement, seating arrangement, and the certain degree of formality that prevails. Heavy traffic, interested bystanders, noise, and curious children can interfere with outdoor training activities. The instructor has little control over these distractions. The instructor usually can achieve control only through preparation and proper motivation of the students. The instructor must be prepared to be in complete charge of the situation at all times.

The use of notes during field training is usually difficult for both instructors and students. Instructors who must refer to notes will usually find that index cards are more suitable than paper. They are easier to hold, do not blow around, and last longer. In general, students should not be expected to take extensive notes in the field. Information that requires note taking can be presented more effectively in a classroom.

As the size of a training class increases, so does the number of students who will have difficulty seeing and hearing the lesson (Figure 6.5). The best way for the instructor to deal with this

5/94

Figure 6.5 Students in large training classes may have difficulty seeing and hearing the lesson.

problem is through proper placement of instructional aids and proper grouping of the students. When the equipment cannot be moved, move the students. The instructor may also use such items as platforms for equipment displays and demonstrations, a portable public address system, and large visual aids.

Weather conditions can have a tremendous influence on the learning environment during field-training activities. Rain, extreme temperatures, and high winds tend to make the instructor's job much more difficult. The instructor should be prepared to deal with weather conditions by moving the group inside, providing frequent breaks, or taking other steps to provide comfort.

ATTITUDINAL SETTING

The physical setting may be just right, but without the proper attitudinal setting established by the instructor learning may not occur. An enthusiastic instructor who sets a positive tone is the key to ensuring that learning occurs. The instructor should serve as a positive role model for the students and should bring to the classroom a sense of professionalism.

This type of instructor will be on time, if not early, to prepare the classroom; neat and professional in appearance; cheerful and friendly; organized; complimentary and supportive of the department; and always encourage students even when difficult material is being covered.

These fundamental principles of setting a positive environment can enhance learning and provide a guide for professional behavior. On the other hand, the attitudinal setting can be negatively affected by an instructor. Negative impacts can include an instructor who arrives late and disregards the time schedule; is disorganized; complains about the department's lack of facilities, equipment, or support services; or is not interested in students learning and performing well, then is impatient with them.

PRINCIPLES OF INSTRUCTION

The instructor's primary responsibility is to make learning as effective as possible for the students. Lessons should be worthwhile and presented so the subject matter is easily understood and meaningful to the class.

Although the principles of instruction are examined from the instructor's point of view, the instructor should not be as concerned with making *teaching* easy as with making *learning* as easy as possible for the student. This is called the student-centered approach. Using this approach, the instructor begins at a level the individual is able to understand, develops new ideas from background knowledge, and leads to a point from where the student can proceed alone (Figure 6.6). Individualized instruction stresses diagnostic and prescriptive instruction to help the student meet the competency. The student-centered approach deals with the whole person.

Certain fundamental guidelines must be followed when using the student-centered method. These guidelines are called the *principles of instruction*. Not all of these principles or techniques can be used at all times, but the instructor should try to incorporate them into teaching plans as often as possible.

Figure 6.6 The instructor should build on the knowledge the student already has.

The First Principle: Start At The Level Of Student Understanding

There are several steps the instructor can take before planning the lesson to ensure beginning at the students' knowledge level:

- Examine the course content to determine what material the students should have already learned.
- Make note of their educational background, experience, age, language, culture, and disabilities.
- Organize lesson material in logical order — proceed from easy to difficult, known to unknown, general to specific.

When presenting theoretical material, the instructor should divide the material into logical stages and present each stage at a rate understandable to the students. The instructor must make sure students have learned one stage before moving on to the next stage.

The instructor may ask review questions at the beginning of a lesson to ensure that students are at the required level. There should be a continuous flow of questions from the instructor to the students and also from the students to the instructor (Figure 6.7). This feedback will aid understanding and will help the instructor gauge how thoroughly the class is grasping the lesson.

In determining the overall level at which to teach the subject, it is advisable to pace the rate of instruction to the majority of the class. Slower students may need individual coaching after class; faster individuals can be given additional work so the lesson is worthwhile to them. An instructor should keep a constant watch for unusual expressions on the students' faces. A raised eyebrow or blank expression may indicate that something is not understood.

Students can be assigned homework covering lessons just completed. A review of the homework is often a good way of starting the next day's lesson. Consistently poor results on assignments may require remedial action by the instructor.

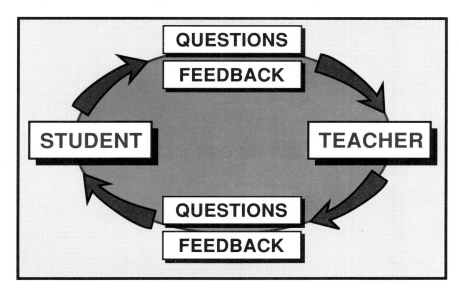

Figure 6.7 Questions should flow from the instructor to the students and from the students to the instructor.

The Second Principle: Emphasize And Support Teaching Points

Clearly worded objectives provide a solid foundation on which to practice this principle. Vital parts of the lesson are readily identified and can be given the emphasis they deserve.

Total lesson time must be allocated relative to the importance of teaching points. It is a good idea to assemble the points in order of importance and then plan lesson times accordingly. If a skill is being taught, make sure there is sufficient time for student practice. Do not spend too much time explaining and demonstrating. Get students involved as early and as often as possible (Figure 6.8 on next page).

Figure 6.8 Students should get involved and practice skills.

Repetition and oral emphasis are both effective techniques of stressing important points. Material presented by the instructor should be substantiated by suitable examples, clear comparisons, meaningful statistics, and quotations from authorities. The following guidelines can also make teaching more effective.

- Use appropriate training aids. Appeal to as many of the five senses as possible.

- Tell the class to take notes during the lesson or give them a well-written handout that contains the highlights of the lesson.

- Assignments that emphasize the main points may take the place of note taking or become notes for the students.

- Conduct reviews emphasizing important points during class.

- Stress vital points by teaching the material step-by-step.

- Plan periodic reviews if the course is long.

The Third Principle: Create And Maintain Student Interest

Instructors should begin their sessions positively and enthusiastically. The instructor who is enthusiastic about the subject will keep the class interested. Throughout the lesson, be alert for signs of boredom or lagging interest. The facial expressions of the students indicate whether they are paying attention. If boredom seems prevalent, it may be time for a break. Changing the physical climate and varying the presentation method may also be necessary.

One way to stimulate interest is to move away from the classroom environment (Figure 6.9). Consult the schedule and

Figure 6.9 Moving away from the classroom environment can stimulate students' interest.

see if there are certain units that can be taught under realistic conditions. If this is the case, take the class outside or to laboratories and let them see an on-the-job situation and use equipment.

The instructor should try to arouse the students' curiosity. This may be accomplished by explaining the purpose of the lesson and stressing the advantages that the new knowledge will provide. The instructor might also plan speed or ability competitions to keep the students' interest. However, the instructor should be careful not to introduce competition too soon, since frustration and lack of interest may result.

The use of more than one instructor for some lessons should also be considered. Students get tired of familiar voices and faces. For this reason, other instructors can sometimes be called upon to demonstrate a technique or help in some other way.

The Fourth Principle: Provide For A Sense Of Success In The Student

Students should know the level of achievement expected during the lesson and also receive a feeling of success during achievement.

There are several ways an instructor can help the student attain a sense of success. Organizing the lesson material in a logical sequence and using clearly worded explanations are two ways to help students learn the subject and consequently attain a feeling of satisfaction.

A valuable tool often overlooked by instructors is the progress chart (Figure 6.10 on next page). Progress charts are generally designed for a specific purpose and, depending upon design, the chart can be either an instructional or administrative aid.

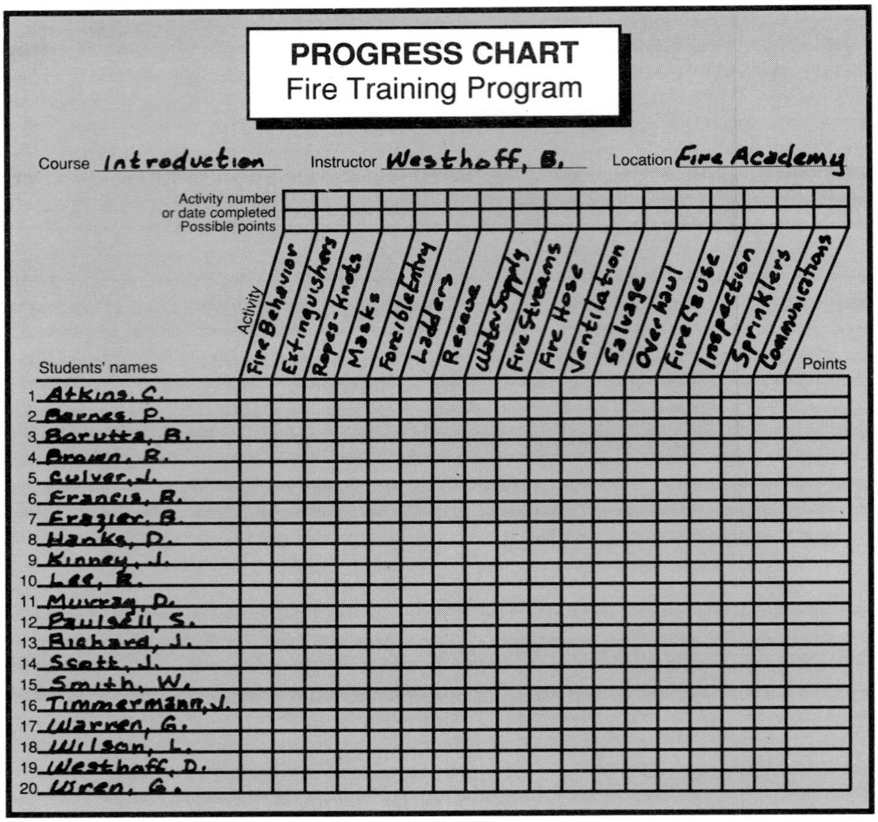

Figure 6.10 A sample progress chart.

When used as an administrative aid, the information arranged and displayed can show an accurate and complete picture of all class activities and work accomplished by both the instructor and students. The chart will also show how the class or program is progressing with respect to calendar or time schedules.

When used as an instructional aid, the information displayed can motivate students by indicating where the student stands in comparison with other class members. Distribution of scores or grades can also be used to identify both strengths and weaknesses in instruction and in learning. The instructor must be careful when information is recorded on progress charts that are to be posted and viewed by all members of the class. Names and grading, or scoring information, cannot be displayed together under provision of the Family Educational Rights and Privacy Act of 1974. However, a distribution of grades and scores may be posted and individuals knowing their grade or score can determine their standing in comparison to other members of the class.

Entries made on a progress chart may include a mark of some type indicating completed assignments, late assignments, additional or outside work, and so forth. Progress charts make the job of the instructor easier and can aid students by providing information about what is expected of them and where they stand.

The instructor should also keep students informed of their progress and praise them for good work. A student who is failing needs correction. The instructor should try to find out why the student is not progressing and correct the situation, if possible.

The Fifth Principle: Provide Meaningful Participation

The instructor should encourage students to answer questions by reasoning with them and develop as much of the lesson as possible. The instructor should ask thought-provoking questions to encourage independent thought.

If lessons involve a skill, plan to have as many students working as possible. If equipment is required for practice, try to obtain one set per student. If this is not possible, the instructor could have some students performing while others identify their actions (Figure 6.11). A beneficial side effect of this type of participation is that it develops confidence in the more retiring individuals by encouraging them to be expressive.

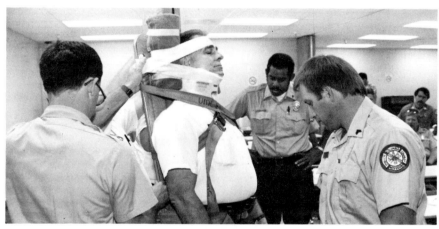

Figure 6.11 Students not performing a task should watch and identify actions.

Remember that any type of participation must have meaning to the students. Stress the importance of doing it right the first time and ensure that the activity is appropriate to what is being learned. Mistakes must be identified and corrected when they are made. People learn from their mistakes only if they are able to recognize and correct them.

For some procedures, the instructor may choose to state a fact or demonstrate a technique while the class observes. The instructor can then call upon the students to explain what they have observed and give examples of how the material may be applied.

The Sixth Priniciple: Reinforce Learning

Reinforcement means "to strengthen or to establish more firmly." A fact heard once may not stick; however, the fact will last for some time if it is heard three times and then written down.

A student's knowledge of facts can be confirmed by questions, problems, quizzes, and written tests. The ability to perform skills must be tested by requiring the student to demonstrate the specific skill. Do not ask a question of an individual to determine ability. The student may be able to tell how to perform a skill but be unable to perform it correctly.

METHODS OF INSTRUCTION

A method is a procedure or manner of doing something. Instructional methods are ways or means of instructing others. A fire service instructor should combine as many methods as are appropriate. Even experienced instructors frequently fail to present subject matter effectively because they neglect certain simple, yet extremely important instructional methods. An instructor will benefit by properly preparing the presentation (Figure 6.12).

Figure 6.12 Properly preparing the lesson will help the instructor present subject matter more effectively.

One way to determine the methods of instruction is based upon what learning is to take place. Therefore, a clear understanding of the lesson objectives within each domain is helpful in the selection of appropriate methods.

For example, a lecture is a good technique to convey new information. Brainstorming is ineffective for this purpose, because students do not have previous knowledge to offer. A demonstration by the instructor is effective for teaching psychomotor skills. It is not appropriate for teaching definitions of terms. A match between instructional methods and the lesson behavioral objectives is a critical function of presenting the instruction.

Methods of instruction vary based upon the amount of combined participation by the instructor and the students in the class. The level of participation is primarily determined by how much knowledge or skill the students bring with them to the class. Their level of knowledge or skill can range from limited to expert. Corresponding to the level of knowledge or skill is the range of instructional methods. In the lecture, the instructor is the only one contributing to the instruction; in a discussion, the students are primarily contributing their knowledge or skills with others in the class. In this case, the instructor serves as a facilitator rather than a contributor.

The Lecture Method

The lecture method of teaching is simply using words to convey ideas. The lecture is often overused because it is the easiest method of instruction. Since lecturing assumes that all students progress at the same rate and because it permits little feedback or application on the students' part, it is often the least effective method of instruction. When using the lecture method, instructors often find themselves telling, not teaching. Proper instruction is based on the exchange of ideas between instructor and student. Therefore, instructors must include some way of getting feedback from the students during the presentation. The lecture can be used effectively, however, to impart material to a large audience in a short period of time (Figure 6.13).

Figure 6.13 The lecture method can be used to convey information to a large audience in a short amount of time.

The greatest cause of failure of a lecture is the mistaken belief that if the instructor knows the subject well, it is not necessary to prepare. The instructor must outline the presentation as thoroughly as if the material were new. This will provide the organization and continuity necessary to cover the subject thoroughly. The quality of the presentation depends upon the ability of the instructor to organize the material and skillfully apply teaching techniques. It can be made more effective with supplementary visual materials. An outstanding instructor will develop new aids, materials, and techniques to make the presentation more interesting.

The initial step in any presentation is for the instructor to gain the respect of the class and favorable attention for the subject. Some of the factors in successful lecture teaching are the instructor's voice, posture, and clothing (Figure 6.14). The first impression that students receive of an instructor is lasting.

Lectures should be delivered in a normal speaking style. This places the students at ease and they will concentrate on what is being said, instead of how it is being said. An instructor should speak plainly and loudly enough for the entire class to hear and should strive for variety in both tone and pitch to fit the subject matter. A slight pause is appropriate and effective to give students time to comprehend difficult material and to take notes. Every lesson or major topic should be summarized.

Figure 6.14 The instructor's voice, posture, and clothing are important when teaching.

The Illustration Method

The illustration method of teaching is a "showing" method since it uses the sense of sight. Showing by illustration includes the use of drawings, pictures, slides, transparencies, film, models, and other visual aids that may clarify details or processes (Figure 6.15). Some instructors are so enthusiastic about illustrations that they sometimes attempt to use these aids as substitutes for a demonstration. Although illustrations can be used to supplement a demonstration, they can never take its place.

Figure 6.15 The illustration method uses visual aids to help clarify instruction.

An instructor who is alert to the possibilities of improving teaching methods will develop other ways in which illustrative materials can be used to effectively aid teaching. All illustrations have certain inherent weaknesses that cannot be overcome if they are used as a substitute for a demonstration. When pictures, charts, diagrams, models, or mock-ups are used as teaching aids, the following suggestions should be considered:

- The drawing or model should be designed to illustrate exactly what the instructor wishes to show.

- Illustrative materials should be well prepared and large enough to be seen easily.

- Keep illustrative materials out of sight until the proper time for their display, unless the instructor wishes them to be seen to arouse curiosity.

- When using charts or drawings to show a series of steps or operations, show only one chart at a time.

The Demonstration Method

The demonstration method is a basic means for introducing new skills. A demonstration is the act of showing someone how to do something. The instructor gives the demonstration while explaining how and why the skills are performed in a certain manner. Sight, rather than hearing, is the primary communication sense. When students get to practice the demonstration, the sense of touch is added to the learning process.

This method is useful in teaching manipulative skills, physical principles, and the working of mechanical devices. Another purpose of a demonstration is to compare products, equipment, and the results of their use. Teaching manipulative skills is more effective using step-by-step demonstrations than by any other means. As simple as the demonstration method appears to be, it can fall short of its goal unless the following guidelines are observed:

- The instructor should have clearly in mind what is to be demonstrated and know where the demonstration will begin and end.

- All fire apparatus, appliances, accessories, and other working materials should be properly arranged and in working order before the demonstration begins.

- The instructor should practice all demonstrations before presenting them to the class. If more than one instructor is demonstrating, the instructors should practice the procedure together. Then in class the procedure will be shown in a consistent and standard way.

- Arrangements should be made for all members of the class to see and hear the entire demonstration.

- An instructor should begin a demonstration by linking new information with the students' knowledge.

- If a demonstration is to be done well, it must be done slowly and with a high degree of skill to emphasize key points.

PREPARATION FOR THE DEMONSTRATION

Planning and preparation are required for a successful demonstration. Considerations in using a demonstration include the size and arrangement of the group, whether the demonstration is to be conducted in a classroom or in the field, and the complexity or accuracy required.

The training aids used during a demonstration can be very distracting to the class if placed on display before they are needed. Instructors should organize and be familiar with all of the tools and equipment in the order in which they will be discussed; further, they should practice several times to ensure that all the materials needed are present and operable. Each piece of equipment should be brought into view when required and removed when no longer a part of the demonstration.

Good planning and constant attention must be given to the arrangement of students so everyone can see the object being demonstrated. Classrooms in which demonstrations will be held should permit variable seating arrangements for better viewing. When a demonstration involves moving the members of the class to positions around a piece of equipment, the instructor should be sure that all students can see by having the class form a semicircle or by having those who press forward move back (Figure 6.16). To prevent obscuring the view of others in the rear, a piece of rope or hose or a chalk mark on the floor provides an excellent boundary for students who are standing.

Figure 6.16 The instructor should make sure that all students can see during a demonstration.

CONTENT OF THE DEMONSTRATION

Every demonstration should contain some theory and background information. The instructor must remember that the basic purpose is to teach through showing. Key points should be covered during the demonstration, carefully explaining how and why the job is done and stressing safety as the operations are performed.

If there are several acceptable methods of performing a particular act, they should be taught, but not in the same class or at the same time. Give the students an opportunity to firmly grasp and practice each movement before introducing the next step or another way of performing it.

SAFETY

Safety must be a part of every demonstration. The best way of accomplishing any job is the safest way. The instructor should stress safety as each step is demonstrated. This is especially important in the fire service because of the hazardous nature of many maneuvers and environments. Some instructors may try to impress the class with speed and agility in performing certain functions and create the mistaken belief that safety should be disregarded. Speed will come naturally and safely as a result of practice when a high standard of safety and performance is maintained.

PRESENTATION

The following sequence of steps is a basic format for a demonstration. Reasons for each step should be explained to develop sequence and fix the step in the students' minds. Questions should be encouraged throughout the presentation.

Step 1: Introduce the lesson; explain why it is important and how it will be taught. Show how the subject fits into the overall program. Proceed from the known to the unknown and from the simple to the complex.

Step 2: Go through the entire operation once at normal speed to give the students a general impression of what is to be learned. Then, repeat the operation slowly, explaining each step. Emphasize the key points in each step.

Step 3: Perform the operation slowly while a student explains each step as it is being performed.

Step 4: Have a student perform the operation and explain each step before it is performed.

Step 5: Have each student practice the operation under supervision until proficient.

One of an instructor's most valuable evaluation techniques is asking questions. The questions should be asked so each individual in the class will be expected to answer them. Not only can the instructor evaluate the success of the presentation through questions, but questions can also be used to emphasize important points. Observation is a critical part of an evolution. The instructor must provide time for and encourage questions from the class, and must respond to every serious question as though it were asked by the entire class. If one student fails to understand, it is possible that the point was vague to other members of the class.

Discussion Method

The discussion method can be effectively used when the students have sufficient knowledge and experience to contribute to the discussion. Discussions of this type deal with available knowledge, ideas, and attitudes of the group and are not suited for passing along new information or skills. The exception is what may develop from a synthesis of the ideas presented by the group.

During a discussion, the burden of developing new ideas or methods is placed on the student while the instructor acts as a coordinator and guides the discussion toward a predetermined goal. A good discussion session will arouse interest as it challenges the students to think and participate. There are several types of discussion methods: guided discussion, conference, case study, role playing, and brainstorming (Figure 6.17).

GUIDED DISCUSSION

An advantage of the guided discussion is that there is more opportunity for personal interaction between the instructor and students than in any other method of instruction. The instructor has instant feedback and a chance to correct misconceptions immediately and tactfully in conversation and without embarrassment. The instructor can assess how much the student has learned and retained from previous lectures by asking leading, probing questions to reveal needed information.

The guided discussion is not a bull session. It must be an orderly exchange of ideas controlled by the instructor. This requires greater resourcefulness on the part of the instructor than do other methods. A reasonable objective for the session must be established and kept in mind throughout the discussion to ensure that all material contributes to the topic.

The participants must be permitted maximum freedom while the discussion is directed toward a predetermined goal. During the discussion, the instructor must constantly analyze what is being said to understand the thinking of the student.

CONFERENCE

The conference method of discussion is used to direct group thinking toward the solution of a common problem. The conference must have a clearly stated end result as a goal. This method has been found to be effective in bringing about changes in thinking and attitudes among participants.

Each member has an opportunity to compare experiences, techniques, and beliefs with those of the group. When one hears the interpretation others have made of a situation similar to one's own, that person is often willing to adopt the attitude taken by the majority of the group.

Conferences conducted on the spur of the moment are seldom a success. Planning by the leader and preparation by the members is important. Sufficient time must be available to permit systematic discussion of all facets of the problem.

All members of a conference group should be willing to share their ideas and trust that the conference results will be better than the ideas of any one member. If one or more of the members, or the leader, is determined to influence the group with personal ideas, the conference will be a failure.

The responsibility for the success of a conference rests upon the instructor. The instructor can greatly influence the participa-

TYPES OF DISCUSSION
- GUIDED DISCUSSION
- CONFERENCE
- CASE STUDY
- ROLE PLAYING
- BRAINSTORMING

Figure 6.17 Discussion methods can arouse student interest and stimulate student participation.

tion of students as well as the direction and outcome of the discussion. Before the conference, the instructor must plan the meeting, provide the agenda, and set the time and place. During the conference, the instructor's function is to introduce and define the problem to be discussed. The primary objective of the instructor is to develop understanding and acceptance of the problem by briefly outlining the background, limits, and scope of the conference and its effect on the decision that will be reached.

Other responsibilities of the instructor are to eliminate bickering and irrelevant discourse, reconcile divergent interests, and unify all participants. It must be clearly understood that on accepting the role of conference leader the instructor is no longer a teacher. The instructor does not tell the group how to think, what to think, or dictate the results of their thinking. While functioning as the conference leader, the instructor must not enter into the discussion except to state or restate problems, ask questions, state cases, or summarize the discussion.

CASE STUDY

In the case study method, the participants review and discuss detailed accounts of past occurrences. This method can be used to develop the students' ability to analyze a situation and examine all the facts necessary to reach a conclusion. It gives students an opportunity to formulate ideas and to discuss their findings with other students.

The instructor must create a classroom or field environment that will encourage the students to review the case and express their ideas without fear of ridicule. Although the instructor may not agree with what a student says, the student must be allowed to complete the presentation. Emphasis is placed more on the reasoning used by the student to reach a conclusion than on the conclusion itself.

Opportunity for case studies arises out of every emergency call. The debriefing or critique is a good place for on-the-job training.

ROLE PLAYING

The role-playing method requires considerable instructor preparation. The instructor must be sure the scenarios are pertinent to the course material. The scenarios should also be clearly defined and capable of being acted out with little difficulty. This can be effectively done by providing each student with a written description of that student's role. The student should not know the particulars of the other students' roles. It is helpful if the students are familiar with one another and relaxed enough to permit active participation within the assigned role.

At the conclusion of the role-playing exercise, the instructor should lead a group discussion that summarizes the results of the activity. Emphasis should be placed on the relationships that developed among the role-playing members.

BRAINSTORMING

Brainstorming can be an effective discussion method if the individuals involved have an adequate knowledge of the subject matter being discussed. During brainstorming, individuals present and discuss ideas in a group. Not all the ideas presented will be workable, but do not be critical of any of them. The purpose of brainstorming is to allow individuals to think creatively within a group. One individual's suggestion may spark an idea for another person, thus eventually enabling the group to come up with a workable solution to a problem. When using brainstorming as a method of discussion, try to get all members of the group involved in the discussion and presenting ideas.

QUESTIONING TECHNIQUES

One of the most important skills required of an instructor is that of questioning. Asking questions that promote class discussion is a useful method of obtaining feedback and communicating ideas to students. The most effective questions are those that cannot be answered by "yes" or "no." The who, what, when, why, where, and how kinds of questions that require a qualified answer are preferred.

There are four general types of questions that can be effectively used to stimulate interest, develop understanding, apply information, and test for evaluation purposes:

- Direct questions
  - — Are addressed to one person
  - — Get the students to express themselves
- Overhead questions
  - — Are addressed to the entire group
  - — Are used to promote thinking, start discussion, and bring out different opinions
  - — Can be responded to by anyone in the class
- Rhetorical questions
  - — Are addressed to the entire group
  - — Are used to promote thinking and to set a general theme
  - — Are not intended to produce an oral response
- Relay questions

— Upon receiving a question from a class member, the instructor relays it as a direct question to another class member

— Can often result in variable information being offered by members of the class

When asking questions, the instructor should display an attitude that is natural, friendly, and conversational. The questions should be within the students' experience and knowledge and should be asked in such a way that the answer is not suggested. Present questions to the class before calling upon someone to answer. Be sure to allow time for thought and for the answer or for a part of the answer.

INDIVIDUALIZED INSTRUCTION

Individualized instruction, although not new in the educational world, has taken a front seat in recent years. The reason individualized instruction has gained popularity is that its approach includes the six principles of instruction.

Individualized instruction is the process of matching instructional methods with learning objectives and students' learning styles to enable a student to achieve the objectives (Figure 6.18). It is a tool for the instructor to use in combination with other traditional methods. It is not a substitute or replacement for the instructor.

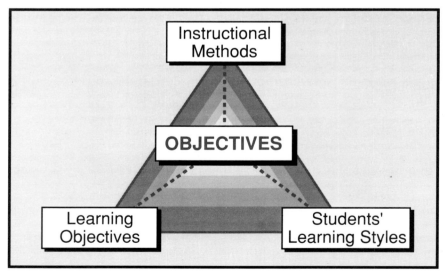

Figure 6.18 Individualized instruction matches instructional methods, learning objectives, and students' learning styles to achieve the objectives.

The process of matching instructional methods with learning objectives creates four major variations of individualized instruction. Learning objectives are typically prescribed by the instructor, but there are times when it can be effective to allow the student to set the learning objectives. Objectives set by the instructor are fixed. Objectives set by the student are optional.

In the same way, instructional methods and techniques are usually dictated by the instructor, but the student may at times specify preferred instructional methods or techniques to accomplish the learning objectives. Likewise, methods and techniques chosen by the instructor are fixed, while those selected by the student are optional.

Individualized instruction is built on three premises: the student's needs and preferred learning style, the learning objectives or competencies required by the occupation, and the instructional strategies and media that fit the needs of the student. The characteristics of individualized learning grow out of these three premises.

Individualized instruction, therefore, is student-centered in that it relates to the individual learner, not the group or class. It is competency-based in that it is based upon the demands of the occupation. In the fire service specifically, the NFPA standards have been set as competencies within the various occupations. Individualized instruction is evaluation-based in that it provides a system for frequent review of progress by diagnosing, prescribing, and evaluating student achievement. It is individually paced, and therefore, flexible in terms of time taken to learn the objectives and in terms of the student's own goals. This flexibility may extend not only to the variety of methods and media used, but also to the manner of enrollment in open-entry, open-exit programs.

Frequent student-teacher contact enables the instructor to facilitate learning by periodically assessing progress, prescribing new learning objectives or different methods used to learn the objectives, and by encouraging students. The role of the instructor in individualized instruction is different from the traditional role. In individualized instruction, the instructor becomes a manager of learning resources, guiding and interacting with students — not the sole or primary resource as traditionally viewed. It is neither easier nor harder to work in the new role. It is simply different and requires different skills than are needed for the traditional "lecture-approach" method of teaching.

Individualized instruction exists in many forms. Some forms that may be familiar are independent study, using learning activity packets with sequenced activities for progressive learning, or using special reading assignments. Tutorial instruction is the one-on-one relationship of a teacher or other class member helping the student. Programmed learning, another example of individualized instruction, is a systematic process of introducing information in small sequential steps, followed by questions designed to reinforce learning. Computer-aided instruction is probably the most fashionable form of individualized instruction because of the increased use of computers. Chapter 9 of this text provides an expanded view of computers in fire training.

SUMMARY

The fire service instructor has the responsibility to present information to students in a manner that encourages and facilitates learning. The instructor should be aware of both the physical setting and the attitudinal setting of the learning environment. The student-centered approach to teaching focuses on making learning as easy as possible for the student. The instructor should be familiar with the six principles of instruction for the student-centered approach and incorporate them into teaching when possible.

The instructor has many choices when deciding how to present information to students. The instructor should be familiar with the different methods of instruction and know when to apply them. The different methods of instruction include lecture, illustration, demonstration, and discussion. The discussion method of instruction can be broken down into guided discussion, conference, case study, role playing, and brainstorming. The instructor may also be able to use individualized instruction.

SUPPLEMENTAL READINGS

Bullough, Robert V. *Multi-Image Media.* The Instructional Media Library; Vol. 9. Englewood Cliffs, New Jersey: Educational Technology Publications, 1981.

Gagne, Robert Mills, Leslie L. Briggs, and Walter W. Wagner. *Principles of Instructional Design.* 3rd ed. New York: Holt, Rinehart & Winston, Inc., 1988.

Gronlund, Norman E. *Individualizing Classroom Instruction.* New York: Macmillan Publishing Company, Inc., 1974 .

7

Training
Aids

This chapter provides information that addresses performance objectives in NFPA 1401, *Fire Service Instructor Professional Qualifications* (1987), particularly those referenced in the following sections:

NFPA 1401

Fire Service Instructor

3-1.1 (l)

3-1.2 (e)(l)

3-6

3-7.1 (f)(g)

4-1 (b)

4-4 (d)

4-5

4-6 (a)

5-1 (b)

5-3.2 (a)(b)(c)(d)(e)(h)(i)(j)

Chapter 7
Training Aids

Training aids are teaching tools that enable an instructor to teach more effectively. Training aids are any material, equipment, or device used to support instruction. Properly selected and used, training aids will help the student better understand and remember what has been taught. Training aids make use of senses other than hearing and reduce the necessity of the instructor to depend on words to carry ideas.

THE NEED FOR TRAINING AIDS

Teaching should appeal to as many of the senses as possible; therefore, the instructor should make frequent use of them. The instructor should carefully analyze each class session and select a variety of training aids that will contribute to learning (Figure 7.1). The actual item being discussed would be the best possible

Figure 7.1 The instructor selects the training aids that best contribute to learning.

aid. If, however, one of the following conditions is present, another training aid is needed:

- The actual item is too complicated, too large, too small, or too spread out to be shown effectively.

- The item or process is not available.

- The item is too dangerous, delicate, or expensive to permit students to practice with it.

- The mechanical movement is too fast for students to see and grasp the movement or the concept.

- The process is not visible to the naked eye.

Training aids are necessary because oral communication alone, even at its best, is inadequate. It is much too easy for conversation to be misunderstood or for the listener to perceive certain words differently than does the speaker. The ultimate difficulty is encountered when attempting to explain an entirely new concept to another person.

Training aids help ensure the clarity of what is being taught (Figure 7.2). For example, a student asked to describe the design of the stairway leading to the top of a lighthouse might call it a spiral. If asked to describe exactly what is meant by "spiral," the student will find it difficult to develop an exact word picture and will probably confuse the audience. Some illustration is needed to transfer the idea of spiral stairs.

Figure 7.2 Training aids help the student better understand what is being taught.

Appeal To The Senses

Training aids help facilitate learning because greater sensory appeal of instructional materials will increase a student's retention rate. The student who can see, touch, taste, or even smell what the instructor is talking about, will have a better and more complete understanding of the information. By appealing to the senses, training aids bring reality and coherence to teaching. What may have taken 10 minutes to develop orally can be summed up with a single drawing or mock-up. This is not to say that with proper aids the lesson will teach itself and an instructor is unnecessary. Training aids are designed to support and clarify instruction — not substitute for it (Figure 7.3).

Figure 7.3 Training aids appeal to the senses and will help increase a student's retention rate.

Saves Time

Time is undoubtedly the most valuable commodity an instructor has, or, as is usually the case, does not have. The instructor must learn to conserve time. Training programs are being squeezed into tighter time schedules. This squeezing takes place both internally and externally. Internal squeezing results when new information as well as fundamentals must be included in lessons; external squeezing occurs because of the increased demand for trained personnel. Consequently, whenever an aid can be used to replace an extended oral explanation, it should be used.

HOW TO SELECT THE APPROPRIATE MEDIA
Behavioral Objectives And Content

Selection of the most appropriate media is based on several criteria. The two primary reasons for choosing a particular training aid are the subject content and the behavioral objectives. The training aids should make the content clearer and the behavioral objectives easier to achieve. A chart showing the different types of tools used in forcible entry will be of little help to the student if the objective is to learn how to vent a roof. A more appropriate aid would be a slide series that shows the procedure step by step.

Class Size And Interaction

The size of the class and the amount of interaction desired will also determine the type of training aid selected. A small class with a lot of interaction could easily use flip charts for discussion ideas. However, in a large class with little opportunity for interaction, an overhead transparency might best be used to introduce a procedure.

Required Student Performance

Different training aids will be used based upon the performance required by the student and the amount of realism needed to convey the information, principle, or procedure. If students must learn how to use self-contained breathing apparatus, then actual breathing apparatus must be used in training (Figure 7.4).

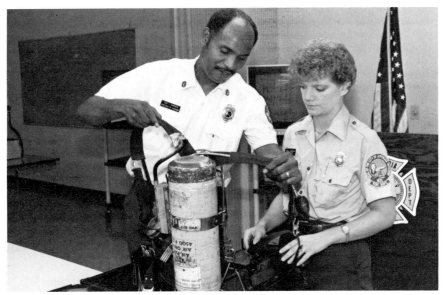

Figure 7.4 Training aids should be appropriate to the lesson being taught.

Flexibility

Other selection criteria are the flexibility of use and the ability to update the information. Some training aids are easier to move and set up than others. Flip charts and easels are usually portable, while movie projectors are not. Some training aids are more convenient to operate. A video is generally easy to operate, but if there are problems, they are not usually easy to fix. The ability to update or edit materials is also an important consideration. A slide series is convenient to update by making new slides and inserting them. On the other hand, a video is a costly production to update.

Pace Of Learning

The ability of a student to have control over the pace of learning can be a determining factor in which training aid to use.

Whereas a movie film gives no control to the student, computer-aided instruction gives the student a great deal of control.

Practical Factors

Practical considerations may make the determination in the selection process. Rarely do instructors have the freedom to consider all the factors just discussed before looking at cost, availability, and preparation time required. Nevertheless, these factors are as important in decision making as any of the others.

HOW TO SUCCESSFULLY USE TRAINING AIDS

Training aids are effective when used properly, but lose their impact when some simple rules are overlooked. Planning and setting up the room in advance is the key to reducing problems during class time.

Seating

A rule of thumb for planning the seating arrangement is that the audience should never be seated closer than twice the width of the screen nor farther away than six times the screen's width. After the classroom is set up, it is wise to sit in different seats around the room to check that vision is not blocked in any way.

Distractions

Another aspect of planning and setting up training aids is to organize and display training aids in the order in which they will be used. Do not clutter the room with unnecessary equipment, tables, or chairs. The learning environment should be free of distractions.

The instructor should ensure that all students can see and/or hear. If the room arrangement or equipment limitations do not permit students to see or hear easily, encourage the students to move. Another option is to use handouts to supplement or replace the visual aids. Remember — do not stand between a student and a visual (Figure 7.5).

Figure 7.5 The instructor should not block the students' view of a visual.

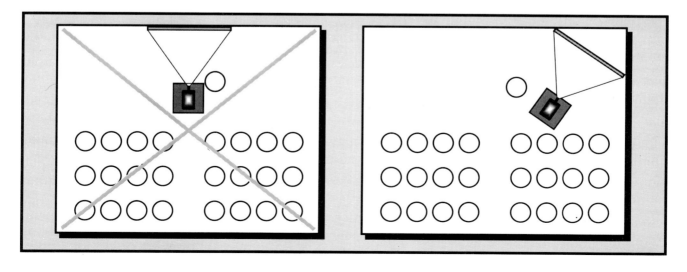

DESIGN CHARACTERISTICS

- Keep training aids simple. An aid should present the essentials without cluttering a presentation with confusing points of information.

- Keep training aids realistic. An aid must be a true picture of an object or activity. The closer the aid is to real life, the better students will be able to understand the information.

- Keep training aids accurate. Avoid using aids that do not provide an accurate picture; this prevents students from acquiring correct information.

- Keep training aids manageable. Select or devise aids that can be easily, quickly, and comfortably managed.

- Keep training aids legible. The size of the picture or the lettering of any visual aid should be large enough to be seen by all students. The size of the picture or lettering should be at least $1/4$ inch (6 mm). An instructor who develops charts and the like should use easily readable, large letters that can be seen from the back of the classroom without difficulty.

TYPES OF TRAINING AIDS

A wide variety of training aids are available. The advantages and disadvantages of the more common types of training aids are described in the following sections. The first section will cover training aids that require that the instructor operate equipment. The second section covers aids that do not require the use of equipment. The final section covers multi-image, multimedia, and simulators.

PROJECTED TRAINING AIDS
Overhead Transparencies

The overhead transparency is made of plastic, cellophane, or acetate sheets with sizes of up to 10 x 10 inches (252 mm by 250 mm). Tracing, drawing, writing, and photographic reproductions made on these sheets can be projected clearly on a screen. With the strong light of the overhead projector, transparencies can be used in an area where blackout is impossible (Figure 7.6).

Additional transparency sheets (overlays) can be superimposed on the base transparency. By using overlays, the instructor can separate processes and complex ideas into elements and present them in step-by-step order. In addition, a felt-tipped pen can be used to add to the transparency during the presentation. If water-based pens are used, the marks can be removed with a soft, damp cloth.

Many commercially prepared transparencies are of excellent quality, but some are not. Often, they are well designed and appropriately colored, but their letters are too small to be legible.

Figure 7.6 The overhead transparency is made of plastic, cellophane, or acetate sheets.

Furthermore, the message may not be designed to specifically support or clarify the teaching point in a particular lesson.

When considering the purchase of transparencies or when developing your own, the following points should be considered:

- Would the subject best be presented on a transparency instead of on a poster, chart, slide, or other medium?

- Is the content up to date?

- Does the transparency make good use of transparency techniques (overlays or other features)?

- Does the transparency meet minimum letter size and color quality?

TRANSPARENCY DEVELOPMENT

Transparencies can be made by using any one or a combination of methods. Among these methods are:

- Handmade, using an acetate sheet and felt-tipped pens or commercially prepared dry transfers. The acetate film used must be compatible with the copying machine selected.

- Machine reproduction using almost any popular copying machine and prepared paper master. (**NOTE:** Any type-writer-produced material should be typed on a machine that has large type.)

Before attempting any method of making a transparency, the instructor should be familiar with the materials, techniques, and apparatus associated with that method. Remember to consult reference books or study the instructions supplied with the various types of materials.

Overhead Projectors

The overhead projector is an effective and easily used visual aid. It is designed to be used in a fully lighted room in front of a class. The image is projected on a screen behind the instructor. The instructor can then face the students when presenting visual information. While facing the class the instructor can write on, point to, or mask off any part of a transparency during its projection.

Instructors should become familiar with some of the terms associated with overhead projectors (Figure 7.7):

Lamp — Light from the lamp shines through the material to the mirror and is then reflected to the screen.

Fan — Controls the inside temperature of the lamp housing.

On/off — Turns on the lamp and the fan. Turns off the switch lamp only. The fan motor is turned off by a thermostatic switch.

Focus Knob — Turning the focus knob, located on the side of the vertical pole, clarifies or sharpens the image.

Mirror — Rotating the mirror will adjust the vertical position of the image on the screen. The size of the image is controlled by moving the projector backward or forward.

Glass — The stage (the horizontal glass surface) of the overhead projector is about 10 x 10 inches (250 mm by 250 mm).

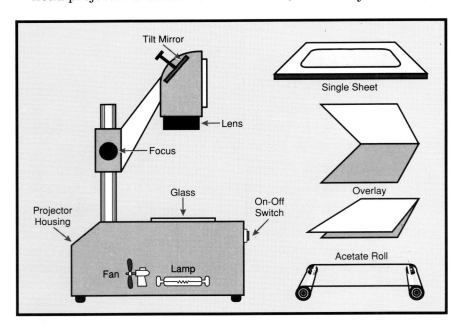

Figure 7.7 Components of an overhead projector.

OPERATION

The overhead projector is operated by following the simple four-step sequence below:

Step 1: Lay the transparency on the surface of the glass. The bottom of the picture on the transparency should be nearest the screen (toward the front of the projector).

Step 2: Turn the switch to "on" and move the projector forward or backward until the projected image covers a suitable area on the screen. To reduce image size, move closer; to increase image size, move further back. At times it may be necessary to adjust the mirror to compensate for the height of the screen. Try to eliminate "keystoning." (**NOTE:** Keystoning is discussed in the next section.)

Step 3: Focus the picture by turning the focus knob.

Step 4: When the transparency presentation is concluded, turn the switch to "off." The fan will keep running until the temperature inside the lamp housing has decreased and then will stop automatically.

KEYSTONING

One of the problems encountered in using a projector is a "keystoned" image. When the projector is not square with the screen, the projected image will be distorted or proportionately larger on one side. In addition, some parts of a badly keystoned image are usually out of focus (Figure 7.8).

Keystoning can be eliminated by tilting the screen until it is square with the projector. Most tripod screens, and some wall screens, are equipped with a device called a "keystone eliminator," that permits the screen to be tilted (Figure 7.9).

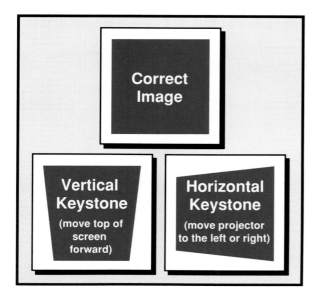

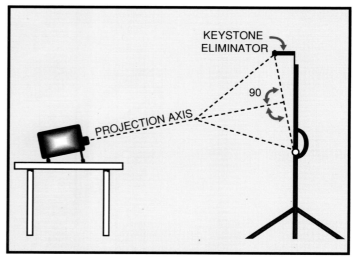

Figure 7.8 One of the problems encountered in using an overhead projector is a "keystoned" image.

Figure 7.9 Keystoning can be eliminated by tilting the screen until it is square with the projector

With an overhead projector the top of the projected image is usually wider than the bottom of the image. There are two ways to correct this: tilt the projector backward or tilt the top of the screen toward the floor, away from the wall.

35 MM Slides

The 35 mm slide is a transparent picture or image on film, mounted in a 2 x 2-inch (50 mm by 50 mm) frame. The image is projected by passing a strong light through it. The slides are compact, convenient to manage, and give a clearly detailed image (Figure 7.10). Slides have great flexibility. The instructor can take slides at various locations, make copies of pictures from books (see the copyright section in Chapter 3), or buy professionally produced slides. Message content can be brought up to date by replacing old slides with new ones.

By showing the slide for as long or as short a time as the instructor wishes, details can be discussed with the students. A general rule is that a single slide should not be exposed to the audience for more than 30 seconds. If the slide contains so much information that a longer period of time is required, the students should be given handouts.

SLIDE PROJECTORS

Slides and slide projectors have become more popular in recent years. Slide projection equipment is simple to use and the skills required to operate it can be learned in minutes. Modern, automatic 2 x 2-inch (50 mm x 50 mm) slide projectors use "trays" that function as storage and as part of the slide feed mechanism. Fully automatic projectors are capable of changing slides at predetermined rates and operate by remote or manual control (Figure 7.11).

Figure 7.10 The 35 mm slide is a transparent picture or image on film mounted in a frame.

Figure 7.11 Slide projection equipment is simple to use and the skills required to operate them can be learned in minutes.

SLIDE PROGRAM DEVELOPMENT

There are a number of systematic ways to go about preparing slide series. For the sake of simplicity, only one method — storyboard — will be discussed here. Whichever method is used, however, there are five preliminary steps to follow before photographs are taken.

Step 1: Identify the subject and write a brief statement about it.

Step 2: Identify the audience. The instructor must take into consideration the background, experience, maturity, preferences, and any special characteristics of the audience. Different groups may require a slightly different approach to a subject.

Step 3: Once the subject and audience are known, specify objectives. Decide as specifically as possible what the audience needs to know about the subject. The instructor should ask: "What do I want to say? To whom? And with what effect?" The instructor may wish to determine what performance or behavior will be expected of the audience after they have seen the program. The learning outcome might be to demonstrate the ability to do a job or to make a passing grade on an examination.

Step 4: Make an outline of the order in which subtopics will be presented to achieve the overall objectives. The outline should be reasonably detailed and should cover the subject thoroughly .

Step 5: Choose the strategy of the presentation. In choosing strategy, the kinds of questions the instructor will need to answer include: How can the selected training aids best be used to achieve the objectives? Should the treatment be light, even comical, or should it be formal? Should the approach be highly personal or impersonal?

When the strategy has been chosen, the instructor is ready to make the storyboard.

With the outline completed and the media and strategy selected, work can begin on the storyboard. A storyboard is nothing more than an array of storycards. Each storycard represents one frame in a slide series (Figure 7.12).

The storycard contains the following four kinds of information:

- Project and frame identification numbers are placed in the upper right hand corner of the card.

- A sketch, contact print, clipping, or some other graphic approximation of what the slide will show is placed in the upper left part of the card.

- The commentary or narration that is to accompany the slide is written across the bottom of the card

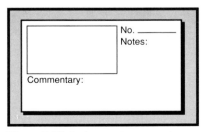

Figure 7.12 A storycard represents one frame in a slide series.

- Any production notes that will help in getting the desired effect when the slides are produced are added to the right center of the card.

Cards of any size may be used for a storyboard, but 5 x 7 or 4 x 6 inch (100 by 150 mm) cards are probably the best choice because they have sufficient room but are still small enough to be handled easily.

Start making the storyboard by writing on each card a summary of what is to be said as each frame is shown. Follow the outline and try to limit each card to only one idea. Often each entry in the outline can be handled with a single card. Occasionally, however, one will need two, three, or even more cards for an outline entry. When possible, try to limit commentary to two minutes or less for a single card.

The card entry need not be the final script material, only a good summary of what is to be said. Identify the project number and frame number for each frame as the work proceeds.

After going through the outline and completing storycards for all the frames, give attention to the illustrations. Choose a scene that will best make the point for each frame. The illustrations can be sketches, photos, or clippings, but each should show what the final slide will look like.

The advantage of using a storyboard in planning is that it allows one to have a preview of the presentation before going to the expense of producing it. Careful study of the storyboard will allow the instructor to avoid mistakes and end up with a better presentation (Figure 7.13).

Figure 7.13 The advantage of using a storyboard is to preview the presentation without going to the expense of producing it.

The storyboard stage of planning is a good time to consider title slides. At this point, one should be able to choose an appropriate title and design both a title storycard and an "end" card.

Filmstrips

A filmstrip is a series of 35 mm single frames processed as a single strip. Each filmstrip is usually from 2 to 6 feet (0.6 m to 2 m) long, with sprocket holes on each side (Figure 7.14). A written or recorded narrative may be included for use with the filmstrip. Since the frames remain in the same sequence, instructors should use filmstrips when the subject matter is of a relatively constant nature.

Figure 7.14 A filmstrip is a series of 35 mm single frames processed as a single strip.

FILMSTRIP PROJECTORS

Filmstrip projectors are designed to project strips of 35 mm film as stills. Filmstrips may have wording printed on them or they may only show diagrams or scenes. They lend themselves to being narrated by an instructor who has complete control over the speed of the program.

Some filmstrips are accompanied by a record or audio tape that gives instructions. This type is ideal when little or no discussion is required and the instruction is relatively specific.

The following are some guidelines for using filmstrips effectively:

- Preview the filmstrip and prepare the discussion.
- Introduce the filmstrip and list points that students should be aware of.
- If the filmstrip is not accompanied by a record or taped discussion, prepare a script.
- Practice until correlation of the narrative and film frames comes easily.
- Tell students what they are expected to get out of the showing.
- Discuss and answer questions throughout the showing.
- Show the strip again, if necessary, to emphasize important points and clarify misconceptions.
- Measure what the students have learned by giving oral, written, and practical tests.

Motion Pictures

Motion pictures can do much to enhance teaching. Firefighter training films dramatize subjects and create a desire in the students to know what is going to happen. Films that deal with a variety of fire protection and related subjects are available from many sources (Figure 7.15).

Many films are not as specific as fire service instructors would like, and for certain limited-audience topics, there are no films available. The instructor must be careful to select a film that will contribute to the lesson plan. When selecting films for classroom presentation, the instructor must be able to answer "yes" to the following questions:

- Does the film contribute to the lesson being studied?
- Does the material suit the experience and educational level of the learners?
- Is it accurate?
- Is it authentic?
- Is it up to date?
- Is the subject matter well organized?

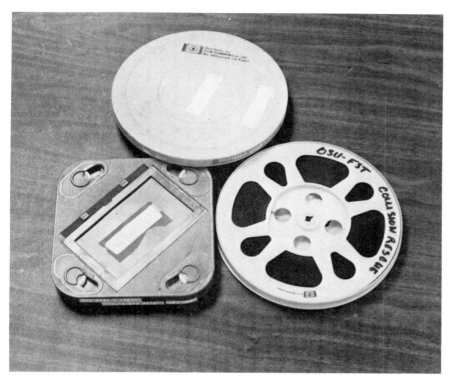

Figure 7.15 Firefighter training films dramatize subjects and stimulate interest in students.

- Does the film benefit the student sufficiently to warrant the time devoted to it? Is the subject matter well organized?

Before showing a film, the instructor should always preview it to determine its teaching content and to become familiar with it. The instructor should take notes during the preview, concentrating on the points that need to be emphasized and on any new terminology or nomenclature that must be defined before the showing.

No teaching method will be entirely successful without class readiness, so the instructor must plan to develop readiness. The instructor should make clear the reasons for showing the film and what is to be learned from it. The following are some preliminary steps that may be helpful in developing class readiness for a film:

- Discuss what the students already know about the subject of the film and lead into what they might expect to learn.

- Introduce key words by listing them on the board. Have the class become familiar with them before the showing.

- Develop a list of questions that are likely to be answered, at least partially, by the film. Put these questions on the board as a guide to viewing.

- Assign certain students to pay particular attention to designated sections of the film, and direct questions to those students during a discussion of the film.

Although a class may have appeared attentive during the showing of a film, the instructor should not assume that the students have learned from the film. After the showing, start a class discussion that relates to the original questions or main points about the film that were established before it was shown. As the students review the film and talk about its content, the instructor can discover any misunderstanding they may have. A second showing, in whole or in part, may be necessary to clear up certain points.

MOTION PICTURE PROJECTORS

The standard motion picture projector used for training has been the 16 mm projector. The "16 mm" refers to the width of the film used. Before becoming proficient at operating a movie projector, the instructor should become familiar with film construction, sound reproduction, and film movement. The instructor should also become acquainted with the electromechanical parts of a movie projector.

The film carries the picture and a variable-area sound track (Figure 7.16). To achieve the illusion of motion, each frame is moved in front of the aperture, held there for a fraction of a second, and then moved out of the aperture, while another is moved in. The movement or transport of the film past the aperture is accomplished by a claw. This claw is a forked device that moves up exactly one frame, is inserted in two sprocket holes, and moves down exactly one frame, all at 24 times per second. As the claw is withdrawn from the sprocket holes it disappears into the projector, only to appear above to move the film down another frame. The shutter and the claw are synchronized. If the film's sprocket holes are not engaged exactly by the claw, the result is evident on the screen as a fluttery picture.

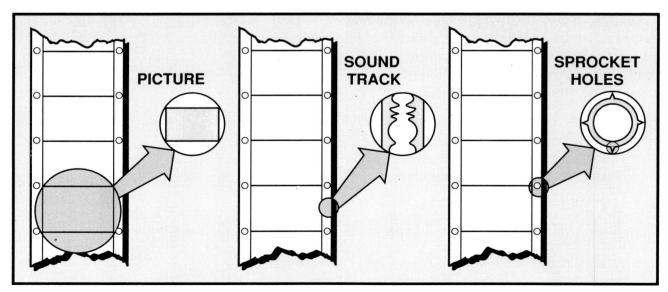

Figure 7.16 Motion picture film carries picture frames and a variable-area sound track.

An important aspect of the movie projector is the sound system (Figure 7.17). There are two methods of producing sounds: optical and magnetic. Most 16 mm movie projectors have optical sound tracks. The optical sound track consists of three components: the sound drum, photoelectric cell, and the exciter lamp. The light from the exciter lamp is intercepted by the sound track on the film. The varying pattern of the sound track is directed against the photoelectric cell, which receives the light impulses and converts them into electrical energy. The electrical energy is amplified and sent to the speaker.

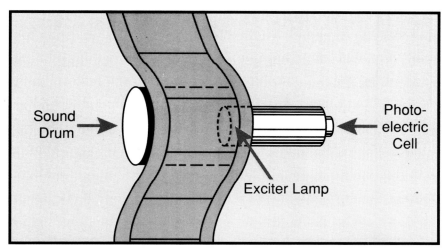

Figure 7.17 Most motion picture sound tracks consist of a sound drum, photoelectric cell, and an exciter lamp.

The film is moved from the supply reel into the threading path and is drawn through the threading path by means of sprockets. Sprockets are wheels with teeth on one side. The sprockets are spaced so they fit the holes in the side of the film. Once in position over the sprockets, the film is then locked into place by clamps or rubber "shoes" that fit over the sprocket.

The sound drum is a heavy roller around which the film must be tightly, or at least snugly, fitted. Film loosely wound around the drum will result in "mushy" sound.

Several roller wheels are used on the projector to guide and direct the film along its path through the projector. These rollers are called snubbers, stabilizers, guides, or pressure wheels by different manufacturers. They keep the film taut and prevent it from rubbing against the projector. Figure 7.18 on the next page is a schematic diagram of the electromechanical parts of a motion picture projector. Depending on the type of movie projector, the motion or film threading will vary.

MOTION PICTURE USE

Planning is necessary before showing a film to a class. The instructor should always preview a film before using it in a lesson

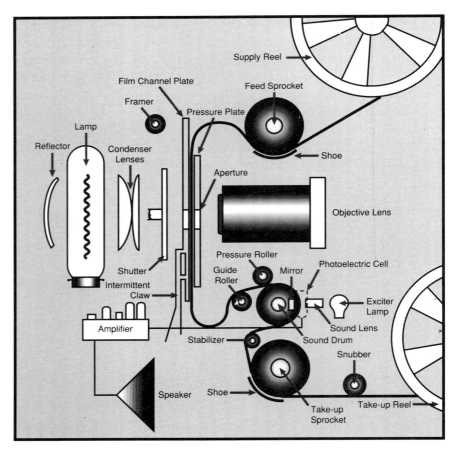

Figure 7.18 Typical schematic diagram of the electromechanical parts of a motion picture projector.

for the first time. By previewing the film, the instructor can introduce the film to the class with the following information:

- What the film will show.
- Why it is being shown.
- What the student will be expected to get out of the showing.

High points — Discuss the high points before showing the film. Emphasize important points the class should remember. The students will recognize these when they see them and the facts will be more likely to make an impression.

Showing — Show the projected material when the students are ready to receive it and when it will help emphasize important points.

Rerun — Show the film again if clarification is needed on unclear points. Experience has proven that retention is quite high after a second or third showing.

Check learning — Hold the students responsible for knowing the content of the film. Check this by testing or discussion.

Review lessons — With a review, clarify any misunderstanding. Another rerun of the film as well as further explanation by the instructor may be needed.

PROJECTION SCREENS

There are several types of projection screen surfaces. Each type of screen surface has different reflective qualities. When selecting a projection screen, give consideration to the room shape and viewing angles. Certain reflection qualities of screen surfaces are more suitable for a particular room setting than others. The following is a brief description of some screen types and their qualities.

Matte Screens — Reflections of light are even in all directions and the images appear almost equally bright at any viewing angle. Viewers should be seated no more than 30 degrees from the lens axis to avoid image distortion. With a matte screen, avoid unneeded light because the screen will reflect it evenly in all directions and degrade the image.

Beaded Screens — These surfaces are covered with small clear glass beads. Most of the light reaching the beads is reflected back toward its source. Therefore, a beaded screen provides a very bright image (up to four times the brightness of that from a matte screen) for viewers seated near the projector beam.

Lenticular Screens — The screen surface is covered with raised patterns that act as small mirrors. These screens reflect nearly all the light from a projector beam over an area 70-90 degrees wide. If not rigid by design, the screen must be kept taut to be effective.

High-Gain Aluminum Screens — This type of screen has a thin sheet of specially grained aluminum foil laminated to a noncollapsible frame. These screen surfaces provide an extremely bright image (about ten times the brightness of a matte screen) over an area 60 degrees wide. Because of the special surface characteristics, the screen can be used in a normally lighted room.

Rear Projection Screens — A rear projection screen should transmit all light from the projector, diffuse the image light evenly to the total viewing area of the screen, and reflect no other light. It should be easy to clean and usually should be rigid. It should not transmit sound, be breakable or scratchable. Since all of these conditions do not exist in one screen, trade-offs will have to be made.

The matte screen is the best for most situations; however, the lenticular screen was developed for wide rooms. The beaded screen is not as desirable because it tends to "yellow" with age and the beads tend to fall off the screen surface. With a beaded screen, the front row should not be as close as with the other screens. Instead of two times the screen width, the front row should be two and one-half times the screen width because of the brightness aspect.

When deciding which type of motion picture screen to select, keep the following points in mind:

- Select a wall screen when portability is not required.
- Check leg locks and height adjustment on tripod screens for ease of operation and substantial construction.
- A keystone eliminator is a must for all types of screens.

Television

The use of television as a training resource has expanded to many different forms. A primary reason for using television is to capture and portray a subject in motion. Television can supply the intricate detail of a process, teach a skill, condense and expand time, and affect attitudes.

LIVE/PRERECORDED

Live programs add spontaneity to the production and in some cases allow for participants to call in answers or questions. Many more programs are now available in a teleconference format on the national level. Live programs are also an excellent way to critique a fire or discuss pre-incident plans. Several drawbacks are the inability to edit inappropriate or incorrect comments and information, and the inflexibility of viewing hours.

Prerecorded programs provide the flexibility of producing the program when most convenient, editing the program, and viewing at any time. This approach, however, does not permit audience participation.

Videotape

Videotapes and monitors can be used in a classroom much in the same manner as a motion picture (Figure 7.19). Programs can be recorded from television and used at a later date. (**Note:** The instructor should be aware of copyright laws regarding recording of programs. See the section on copyright in Chapter 3.) Educational videotapes may also be purchased to aid instruction. Relatively inexpensive, portable, lightweight equipment can be used for training operations both indoors and out. Closed-circuit transmission of videotapes has the following advantages:

- Training is standard.
- Any number of receivers may be attached to the playback unit.
- One instructor can teach vast numbers of students.
- Extremely small items can be magnified and shown clearly to all students.
- A very dangerous item or process can be shown without danger to students.

Figure 7.19 Videotapes and monitors have many benefits and are easy to use.

With a videotape system one may have "instant replay," which can be controlled by the instructor and used for analysis of performance.

A videotape system can be an excellent training tool because it can be used with any television receiver and closed-circuit television. These systems consist of a black and white or color camera, recording/playback unit, and a television monitor (Figure 7.20). Sound can be recorded. Videotape systems are now

Figure 7.20 Videotape systems consist of a camera, recording/playback unit, and a television monitor.

available as compact portable units that can be used at training locations and on the fireground. Several advantages of the videotape system are that prerecorded programs can be developed for later use and that tapes can be erased and used again.

VIDEOTAPE PRODUCTION

Today's viewing audience is accustomed to professional quality. If the quality is missing, the message is often overlooked. For this reason, there should be a script, a subject that lends itself to television as a medium, and talent, whose voice and presence adds to the production.

Script — The script must contain every statement that is to be made on tape. Even professional television personalities use cue cards or teleprompters. A flip chart with large lettering will suffice.

Since a script is the "storyboard" for video productions, it should be in the language of the audience and the language that the talent will use. (See "Storyboard Construction" on page 176.) Do not use words that the talent person is uncomfortable with. Also, by positioning the flip chart or cue cards near the camera or using the teleprompter, the talent person will give the impression of looking at the audience.

Accuracy is a must. Since the production can and undoubtedly will be critiqued, information sources should be checked closely. This should be done during scripting. It is both dangerous and embarrassing to find a mistake during shooting, or worse, while the program is being broadcast.

Audio — If the use of facilities to do voice-overs is available, it is effective to film the visual or action shot and provide the audio portion at a later date. This procedure will provide clear, concise audio text by eliminating such background noise as running saws, pumpers, or traffic.

Studio Time — If present video style has been to take a 1/2-inch (13 mm) camera out to the drill ground and spontaneously create training programs, it is not serving the audience or the subject as it should. Television studio time can be rented to make productions more professional. Some television stations will make their facilities available at reduced rates.

Equipment — The needs and budget of a department will dictate the type of video taping equipment that a department uses. It is advisable to use the largest format affordable. If the department plans to supply television stations with action shots, almost all stations will accept the 3/4-inch (20 mm) format and many stations are now allowing 1/2-inch (13 mm) tapes.

The prices for video equipment differ greatly between format size and manufacturer. As a rule of thumb, ³/₄-inch (20 mm) player/recorders are more expensive than comparable ¹/₂-inch (13 mm) units. Blank ³/₄-inch (20 mm) tapes are also usually more expensive than blank ¹/₂-inch (13 mm) tapes.

The most common format for ¹/₂-inch (13 mm) tapes is VHS. The VHS format is far more common than the BETA format and BETA is now hard to find. The vast majority of tapes sold and rented are VHS.

Quality video cameras have decreased greatly in price since their first introduction but are still a relatively major expense for many fire departments. The needs of the department and extent of equipment needed should be carefully evaluated. Look around before purchasing any equipment.

Video Projectors

With the popularity of videos and computers, equipment to use with videotapes and computer systems has become available. Video projection systems allow users to project video images on a screen (Figure 7.21). There are even video projectors that can be connected to personal computers, allowing large groups to view video/data output without a lengthy delay. Although much of the equipment is relatively expensive, it would be beneficial for organizations that do a lot of presentations. The use of video and computers in education is a new field and is likely to expand in the future.

Figure 7.21 Video data projectors allow video images to be projected onto a screen.

Slide-Tape Machines

A slide-tape machine is a combination slide projector and audio tape player (Figure 7.22). Slide-tape machines may be designed as a single unit with a rear projection screen or as an individual unit capable of projection onto the front of the screen only. Slide-tape machines can be used to develop prerecorded units of instruction and are suitable for self-instruction programs. Single-unit machines are effective for individual and small group instruction of up to five or six persons.

Figure 7.22 A slide-tape machine is a combination slide projector and audio tape player.

AUDIO TAPE PRODUCTION

The two kinds of tape recorders available are reel-to-reel and tape cassette. Either can be used effectively, but the cassette recorder will be discussed here because of its convenience and growing popularity (Figure 7.23).

Most educational tapes run 30 minutes or less. Consequently, a type C-60 (30 minutes on each side) cassette is a good choice in most cases. It is often wise to use a new or completely erased tape for each recording. This practice will eliminate the possibility of having some previously recorded material play in the middle of the tape. To erase a previously recorded tape, it may be necessary to replace the record lockout tab on the rear of the cassette (Figure 7.24).

Notice that on the cassette shown in Figure 7.24, the lockout tab has been removed from the left side but the one on the right is still intact. With this cassette, a recording could be made only on one side (the side associated with the right-hand tab). To record on the other side the lockout hole must be blocked off. This

Figure 7.23 The cassette-type tape recorder is convenient and very popular.

Figure 7.24 Notice that the lockout tab has been removed from the left side, whereas the one on the right is still intact.

can be done by filling the hole with aluminum foil or by covering it with tape. With the lockout tabs removed, there is no danger of erasing the programs.

A detailed script of what is to be taped should be typed double-spaced or triple-spaced in a form as follows:

- The lines should be short so they can be easily read with little eye movement.
- Everything should be written out exactly as it is to be read, using no numerals or symbols.
- Each frame should be clearly set apart from the others.
- Words to be emphasized, pauses, and change-of-frame cues should be marked.

Before attempting to make a recording, rehearse the script aloud several times. Remove any words or phrases that do not flow. The audio and visual materials must be well planned and should present a unified message. When recording a long script, it is often convenient to rehearse and tape it a few frames (scenes) at a time. Any reasonably quiet room will do for recording.

If the instructor chooses to use background music in an instructional recording, it is necessary to get permission if the music is covered by a copyright.

Change-of-frame cues can be added in several ways. If the equipment is available, the use of an inaudible automatic change is desirable. In other instances, an audible automatic manual change cue can be used. The instructor can produce such cues with a tone generator or by stroking a partially filled water glass with a metal spoon.

The following suggestions are given for more effective tape recorder use:

- Set up the recorder according to the manufacturer's instructions.
- Attempt to eliminate all background noise.
- Record the sound.
- Place microphone for best recording quality.
- If the recording is to be used as a part of a presentation, the lesson plan should include an introduction, discussion points, and questions.

NONPROJECTED TRAINING AIDS
Chalkboard
The chalkboard is a convenient and effective visual aid that permits students to take notes while the instructor is listing important instructional points (Figure 7.25). The chalkboard is helpful when stressing significant subject matter. Mathematical and other problems can be solved in front of the students. Illustrations can be easily developed with different-colored chalk.

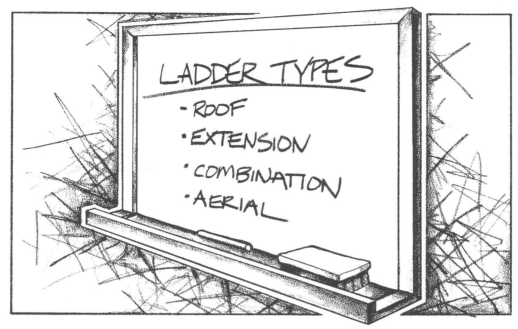

Figure 7.25 The chalkboard is a convenient and effective visual aid that permits the instructor to list important teaching points.

In addition to the traditional chalkboard, there is also the white chalkboard or white-board. These types of chalkboards commonly use felt-tipped markers. There are advantages to these types of chalkboards. There is no chalk dust like there is with traditional chalkboards. There is also a greater range of colors that can be used on white chalkboards. A disadvantage is that these chalkboards can be harder to clean than the traditional ones. The surface of these boards should be very smooth and hard because colors will penetrate. This can make the problem of "ghosting" greater with these devices. Some markers require special erasers or solutions. When using these types of chalkboards, the instructor must be aware of the cleaning methods and proper drawing materials for the chalkboard.

The following are suggestions for effective use of a chalkboard:

- Keep the chalkboard clean.

- Put complex drawings on the board before class time.

- Use a T-square, compass, ruler, and protractor when accuracy is important.

- Make a pattern for frequently used symbols, diagrams, or drawings.

- Present new words, technical terms, definitions, outlines of key points, rules, and important ideas.

- When lecturing, do not stand in front of material written on the chalkboard.

The advantages and disadvantages of chalkboards are as follows:

Advantages

- Readily available and versatile
- Logs progression of discussion
- Keeps students' attention
- Highlights key points
- Requires no special equipment or mechanical skills

Disadvantages

- No permanent record
- Sometimes requires quick thinking
- Requires idea condensing and restatement
- May lack portability

Flip Charts

The flip chart is generally composed of several pages of related material. Each page exposes one or more new points to be viewed by the students. The pages may be flipped as one flips the pages in a notebook.

Flip charts are generally used to accompany and help illustrate verbal commentary. Because of this, flip charts should not be overloaded with words. The effectiveness of flip charts depends on how well the instructor integrates the verbal and visual aspects of the lesson.

Flip charts are one of the easiest and most convenient forms of visual aids to use, with the possible exception of the chalkboard. They are convenient teaching aids because they provide a permanent record and are portable (Figure 7.26).

The instructor can make several different kinds of flip charts. Before the chart is constructed, the instructor must decide which type will best fit the needs of the subject and what exactly is to go on the chart to illustrate the message. Different types of flip charts are discussed in the following paragraphs.

Simple Chart — A simple chart exposes everything on the chart at one time. The simple chart usually consists of words, phrases, and diagrams that can be easily seen and quickly read.

Strip Chart — A strip chart conceals with strips of paper some of the items on the chart. These strips can be removed as the hidden material is needed in the course of the discussion.

Pinup Chart — The pinup chart is built of separate strips that are pinned or attached to a display board.

Figure 7.26 The flip chart is generally composed of several pages of related material.

Window Shade Chart — A white or cream colored window shade makes an excellent rolled chart. It is easily prepared and stored, and will be usable for a long time with minimum care. Sketches or illustrations of items may be progressively revealed as the shade is unrolled.

FLIP CHART CONSTRUCTION

All flip charts should be the same size to make filing simpler and handling easier. Flip charts should be large enough to be clearly viewed. The extent to which the chart will be used or the length of time it will be in service determines the type of material to be used.

Legibility is very important. The instructor should remember that plain letters are more readable than fancy ones. The size, style, spacing, and location are important if lettering is to be legible and helpful to those who view it.

- Black and white lettering or colored letters on sharply contrasting backgrounds are better than combinations with low contrast.

- Letters with a width-to-height ratio of about three to five are better than extremely thin or extremely wide letters.

- Do not try to put too much on one chart. Crowding letters and placing too many items on a chart will reduce effectiveness.

The advantages and disadvantages of flip charts are as follows:

Advantages

- Same as with chalkboard
- Instructor has permanent record
- Advance preparation possible
- Can be reused or revised when needed

Disadvantages

- Preparation and changes usually take time and money
- Less flexible than chalkboard
- Size limits audience
- May be difficult to transport if large

Illustrations

Illustrations can help the instructor explain, clarify, and summarize the lesson. These visuals may include charts, diagrams, posters, and pictures (Figure 7.27). One advantage of illustrations is that they can be easily constructed by the instructor to meet the specific needs of the teaching situation. They are particularly useful when the lesson includes operations in a small area or with small tools, since illustrations can be made large enough for the entire class to see. Illustrations show only the necessary detail. Illustrations should present only what the instructor discusses, not minute details that tend to confuse the students.

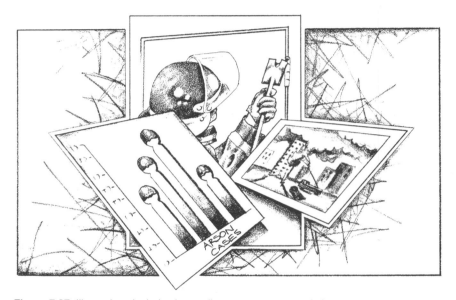

Figure 7.27 Illustrations include charts, diagrams, posters, and pictures.

The advantages and disadvantages of illustrations are as follows:

Advantages

- Materials readily available
- Ideal for study displays
- Can be used in a lighted room
- Can be enlarged to show detail

Disadvantages

- Difficult to store and transport
- Limited to small group instruction
- Original drawings not easily altered

Display Boards

The display board is a popular visual aid since it is effective for portraying illustrations and pertinent information, especially when ideas are to be presented one at a time (Figure 7.28). A display board can be constructed of sheet metal to which small

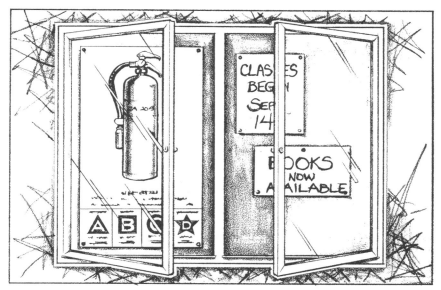

Figure 7.28 The display board is a popular visual aid since it is effective for portraying information, especially one idea at a time.

magnets are attached. A display board can also be made by covering plywood with a piece of felt or hook-and-loop material wrapped tightly around the board and stapled with an ordinary office stapler.

When developing information to go on a display board, make certain proper emphasis is given to important parts by their size and color. Plan the portion of the presentation for which the display board will be used so each piece is introduced as an integral part of the lesson. Stack each piece in the order in which it will be introduced. As the lesson progresses, place each piece firmly and exactly where it fits on the display board.

The advantages and disadvantages of display boards are as follows:

Advantages

- Can be tailored to individual class needs
- Easy to use and produce
- Instructor can arrange and rearrange the material at will

Disadvantages

- Audience limited by size of visual
- Improper use may provide a distraction
- Equipment not readily available

Duplicated Materials

Although training aids such as chalkboards, flip charts, and display boards are versatile and valuable teaching aids, the instructor must take care not to waste class time by having students copy too much from them. One way to avoid this problem is to use duplicated materials as handouts (Figure 7.29).

Modern duplication methods make handouts easy. Anything that can be typed or drawn on a piece of paper can be easily duplicated. Most instructors use these aids for assignments and examinations and to provide students with copies of selected instructional materials.

Figure 7.29 Duplicated materials used as handouts can be used to provide points of information.

The advantages and disadvantages of handouts are as follows:

Advantages

- Helpful as follow-up
- Students can give full attention to instructor rather than to note taking

- Ease of duplication
- No group size limits

Disadvantages

- No assurance material will be read
- Paper and printing costs
- Distribution may disrupt class
- Research and development time needed for preparation

Models

A model is a scale replica of something. Models, particularly working models, can be very effective teaching aids. With a model the student not only sees but can examine the device closely and operate it (Figure 7.30).

Figure 7.30 With a model, the student not only sees but can examine the object closely.

Models vary greatly in form, depending upon the purpose for which they are designed. The following types of models are useful in the classroom.

The Cross-Sectional or Cutaway Model — Cross-sectional or cutaway models make it possible for a student to see the interior of a piece of equipment or machine to understand its inner workings. These are some of the most inexpensive models. The parts may be labeled or the instructor may point out their location while telling their function in relation to the operation as a whole.

The Working Model — Working models are intended to show how a tool or machine operates. Working models may range from simple to very intricate. Although a working model may differ in size and appearance from the original object, the model demonstrates the basic principle of operation. When building a mechanical model, use contrasting colors for the various moving parts. The colors clearly identify and distinguish the various operations.

Transparent Model — The chief value of a transparent model is that a student can see the interior, rear, and sides from a single position. More than any other model, it is important for a transparent model to be accurately built to scale. If there is any inaccuracy, the student will not be able to understand how the component parts fit together or how the machine operates.

The following procedures will help make the use of models more effective:

- Plan the lesson around the model.
- Be sure that the correct concept of size is given.
- Be certain that everyone can see.
- Mount models so they can be moved and conveniently used.
- Have students examine the model after proper instruction has been given.
- Do not pass the model around as it is talked about — it only becomes a distraction.

MULTI-IMAGE
Multi-image refers to more than one image shown at a single time projected by one type of media. For example, a 16-carousel projector program for projecting images used to fill two or more screens or parts of a screen is a multi-image show.

MULTIMEDIA
The multimedia approach uses a combination of instructional materials. These materials, often called instructional kits or learning packages, may include written, pictorial, and audio media. The combinations are limitless, depending on many factors — some personal or some physical, in terms of training facilities and services.

Audiovisual equipment has been developed by manufacturers to facilitate the use of multimedia programs. Some equipment can record student responses and combine automated and manually controlled operation in presenting sound or silent filmstrips, motion pictures, slides, taped narrations, and quizzes.

SIMULATORS
The use of simulators is a rapidly expanding area of teaching because they are very effective for training. Simulators can be

used to create a situation in which students must apply what they know or can experiment with different strategies without the hazard of loss of life and property. Depending upon the type of simulator, different kinds of training can be conducted. Prefire planning, post-fire analysis, company training, fireground status, command and control, search and rescue, and public education are all suitable uses.

Electronic Or Mechanical Simulators

Fire control simulators come in different designs and capabilities. One type is the electronic or mechanical device designed to create realistic fire situations. This type of simulator incorporates the use of slide and overhead projectors with sound-producing and receiving equipment.

To begin the simulation, a photograph of a problem building, facility, or area is projected onto a screen. Personnel who are to participate in the exercise are assigned certain roles to act out. The instructor creates a fire hazard situation on the screen by placing additions on the overhead projectors that create the illusion of fire, smoke, or other factors. The participants then respond by routing and placing equipment, laying lines, and otherwise reacting to solve the problem.

This simulator requires planning by the instructor, but is very effective for applying student learning.

Display Board Simulator

The same exercise can be used on the steel surfaced display board. This design supplies magnetic symbols to mock up fire operations or mass casualty incidents on the steel board. Building outlines and floor plans can be laid out; apparatus, hydrants, and personnel can be located; and hoselines can be placed. The simulator board can also be used to diagram the fire department's organization and chain of command.

The simulation can be preserved by taking a color slide of the board. It can be changed by rearranging the magnetic pieces. This training aid lends itself to group participation and problem solving. It eliminates the mess of a chalkboard. It is very flexible and cost-efficient.

Smoke Simulators

Smoke generators are a very different kind of simulator. This simulator produces a thick but harmless vision-impairing smoke. It is perfectly suited for search and rescue training. It allows students to experience the trapped feeling of being in a live fire, the practice of searching a building, and rescuing a person safely under simulated conditions. The mock search and rescue training is more cost-effective than live fire training and produces the same results. (**Note:** See Appendix B for a description of a number of fire fighting training aids that are simple to construct.)

SUMMARY

The fire service instructor should use training aids to help students better comprehend information. The instructor should be familiar with the different types of training aids and know when to use them. There are many different considerations when deciding which training aid is most appropriate, but practical considerations may limit the instructor.

Training aids include both projected and nonprojected aids. Projected training aids require equipment to operate, while the nonprojected aids do not. Projected training aids include overhead transparencies, 35 mm slides, filmstrips, motion pictures, television, videotapes, slide-tape machines, and tape recorders. Nonprojected aids include chalkboards, flip charts, illustrations, display boards, duplicated materials, and models. Simulators are a different type of training aid and include electronic or mechanical simulators, display board simulators, and smoke simulators.

An effective instructor will use a variety of training aids and bring a certain amount of creativity into the classroom. Although fire service instructors may not have access to all the training aids listed in this chapter, they should use the materials available to them and be innovative when using and designing training aids.

SUPPLEMENTAL READINGS

Anderson, Ronald H. *Selecting and Developing Media for Instruction*. Madison, Wisconsin: American Society for Training & Development, Van Nostrand Reinhold Company, 1976.

Bullough, Robert V. Sr. *Creating Instructional Materials*, 3rd ed. Columbus, Ohio: Charles E. Merrill Publishing Co., 1988.

Gerlach, Vernon S., and Donald P. Ely. *Teaching and Media: A Systematic Approach*. Englewood Cliffs, New Jersey: Prentice-Hall, Inc., 1971.

Kemp, Jerrold E., and Deane K. Dayton. *Planning and Producing Instructional Media*. New York: Harper & Row Publishers, 1985.

Romiszowski, A.J. *The Selection and Use of Instructional Media*. 2nd ed. New York: Nichols Publishing, 1988.

Simonson, Michael R., and Roger P. Volker. *Media Planning and Production*. Columbus, Ohio: Charles E. Merrill Publishing Company, 1984.

8

Evaluation
and Testing

This chapter provides information that addresses performance objectives in NFPA 1401, *Fire Service Instructor Professional Qualifications* (1987), particularly those referenced in the following sections:

NFPA 1401

Fire Service Instructor

3-1.1 (g)(h)

3-1.2 (n)

3-10.1

3-10.2

3-10.3

3-10.4 (c)

3-11

4-1 (c)

4-7

5-1 (d)

5-4.1

5-4.2

5-4.3

5-4.4

Chapter 8
Evaluation and Testing

Evaluation of training methods is as much a part of the instructor's job as is the actual teaching. This job is critical and must not be overlooked by the instructor. Fire departments must have a way of measuring the effectiveness of their training methods. Testing is one way of evaluating how much students have learned.

Fire departments should not base their training methods on what has been done previously. The fire service is constantly changing and training methods should be kept current with what will be used on the fireground.

PURPOSE OF EVALUATION

The primary purpose of evaluation is to improve the teaching/learning process. Many evaluation techniques are available to assess how much the students learned and to determine how instruction can be changed to enhance student learning.

Evaluation is the systematic collection of information for decision making. It has three major components: criteria, evidence, and judgment. For example, in evaluating student learning, the instructor must have well-defined behavioral objectives (criteria), test results or observations (evidence). Based on these, the instructor must be willing to make a decision (judgment) whether the test results (evidence) indicate accomplishment of the behavioral objectives (criteria) set at the beginning of the class (Figure 8.1 on next page).

Evaluation does not occur if one of the three components is missing. As an illustration, if no behavioral objectives exist, there is no way to judge what the test results mean or to compare them against set objectives. On the other hand, if no one is willing to

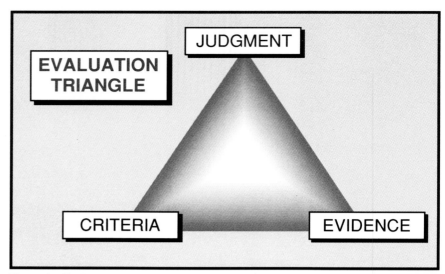

Figure 8.1 Evaluation consists of criteria, evidence, and judgment.

make a decision or judgment about how well the behavioral objectives and the test results match, then no evaluation takes place.

Evaluation is also used for other purposes. It can be used to discover weaknesses in learning and instruction, diagnose the causes, or to stimulate students toward further study. It can also be used to assign grades or to make administrative decisions.

PURPOSE OF TESTING

Testing is an integral part of instruction and evaluation. It is the measurement of learning results (one type of evidence); and therefore, only a part of evaluation (Figure 8.2). The purpose of testing student learning is to determine the degree to which students learned what the instruction intended for them to learn, based on the behavioral objectives. However, test results alone are not evaluation.

Testing is often used as a tool to determine if an employee should be promoted. The organization of the fire service is based on rank. Promotions in the fire service mean an increase in responsibility and in pay. Because of this, fire departments must be very careful when dealing with the promotional aspects of the fire service. There are both legal and personnel considerations the department must take into account. The same consideration given to hiring employees must also be given when promoting those employees.

The guidelines and criteria for promotions must remain uniform for all employees. Fire departments should have their promotion policies in writing and employees should be aware of them. The department should keep detailed records of its promotional activities and follow the guidelines in the Family Education Rights and Privacy Act of 1974 regarding individuals' scores

Figure 8.2 Testing is the measurement of learning results and is one part of evaluation.

on promotional exams. It is not acceptable to make any exceptions to standard promotion policy. Promotions should be made on an individual's capability to handle the position. In the fire service that capability is often demonstrated through testing.

When using tests as a means for promotion, the department must ensure the reliability and validity of the tests. Copies of tests must be closely guarded and preferably locked in a secure place. If an individual has seen a copy of the promotion test, the testing procedure will not be valid.

Criteria for evaluating tests must be established and preferably written in the promotion policy guidelines. If there is a conflict or disagreement with an employee, the department will have written documentation to substantiate its promotion policy. Fire departments should have a standard procedure to deal with grievances regarding promotional testing. A panel of fire department personnel should hear the grievances and settle the matter within the department, if possible. An employee who believes he or she has been unjustly denied a promotion may decide to take legal action against the fire department. If this happens, the department should be prepared to back up its position with written documentation.

CLASSIFICATIONS OF TESTS

Different types of achievement tests can be used at different times during instruction. To choose the type of test that will best measure what is to be measured, the instructor must have a basic knowledge of the various types of tests and their uses. Each type of test needs a specific test design. Therefore, classifications have been designated to help classify the tests for useful purposes.

A classification is merely a way of grouping tests based on a related factor. Within each classification, however, distinctions are made among the various tests. A simple classification helps an instructor distinguish among tests and then select the one that best suits the purpose. Three classifications will be covered in this chapter:

- The first classification is based on the method of interpreting test results. Criterion-referenced and norm-referenced are two methods of interpreting tests under this classification.

- The second classification is based on the purpose of giving tests and the point at which they are given during instruction. Prescriptive, progress, and performance tests are examples of this classification.

- The final classification is based on the method of administration. Oral, manipulative-performance, and written tests are methods of administration found in this classification.

Classified According To Interpretation Method
CRITERION-REFERENCED

Criterion-referenced tests attempt to describe individual student performance. Test results are compared against clearly specified behavioral objectives to determine whether learning has occurred. The emphasis is on meeting a minimum acceptable standard, otherwise known as mastery. Therefore, grades are not stressed, but "mastery" or "nonmastery" are generally used to describe student performance.

Criterion-referenced tests usually focus on a few learning tasks but include all items on the test to accurately describe performance. Test items are not changed to make them more or less difficult; they are designed to measure the exact performance expected.

NORM-REFERENCED

Norm-referenced tests try to discriminate performance levels among students within a class or group. Test results rank students on a comparative basis. The emphasis is on assigning grades that distinguish students from each other.

Norm-referenced tests generally cover a larger area of achievement than do criterion-referenced tests. Test items are chosen that allow maximum discrimination among students. Easy items are discarded and items of average difficulty are favored, with a few difficult items added.

Although criterion- and norm-referencing are different ways of interpreting test results, they can be applied to the same test. It can be said that Mike exceeded 90 percent of the class (norm-referenced interpretation) by calculating 20 of the 25 chemical equations correctly (criterion-referenced interpretation).

Because instruction in the fire service is geared to training at a minimum acceptable standard, instruction and testing of student progress and achievement should be criterion-referenced. With mastery being the goal of instruction, arrangements should be made for remedial activities to bring nonmastery students to the minimum level of learning. As an example, professional qualifications standards are typically criterion-referenced, while promotional examinations are usually norm-referenced.

MANIPULATIVE-PERFORMANCE

Manipulative-performance tests are used to measure an individual's proficiency in performing a job or evolution such as achieving a psychomotor objective. They usually hold the test-taker to either a speed standard (timed performance) or quality standard (minimum acceptable product or process standard), or both. They are the most direct means of finding out how well an individual can do a job. The use and care of tools and equipment, apparatus driving and operation, and emergency care techniques are examples of psychomotor areas for which student performance is evaluated with a manipulative-performance test (Figure 8.3).

Figure 8.3 A manipulative-performance test can be used to measure an individual's proficiency in performing a job.

Manipulative-performance tests should not be confused with drilling. The purpose of drilling is to give learners an opportunity to practice skills. The purpose of manipulative-performance tests is to give learners an opportunity to demonstrate their proficiency under rigidly controlled conditions. Only when such conditions are present can the instructor make valid and reliable judgments regarding student performance.

CONSTRUCTION AND USE OF MANIPULATIVE-PERFORMANCE TESTS

To construct and use manipulative-performance tests the instructor should take the following steps:

Step 1: Specify the performance objectives to be measured, then construct test items based upon those objectives. Each test item should require the performance of a number of basic skills. This will allow a broad sampling without consuming the time necessary to test performance of each basic skill. For example, the instructor might use a test item based upon ventilating a pitched roof that requires use of ground and roof ladders, various cutting tools, safety ropes, and a hoseline.

Step 2: Select the rating factors upon which each test will be judged and design a rating form. Rating factors for performance tests usually include the student's approach to a particular job or procedure; care in handling tools, equipment, and materials; accuracy; and time required to complete a job or procedure safely. A student should be rated against a standard, not against other students (Figure 8.4).

Step 3: Prepare directions that clearly explain the test situation to students. A written set of instructions supplemented with an oral explanation and an opportunity for students to ask questions are necessary if students are to understand what is expected of them.

Step 4: Try out a new manipulative-performance test on other instructors, if possible, before using it on students. A trial test may uncover problems that can be corrected and the test will be more valid.

Step 5: Use more than one rater to increase test reliability. Other instructors or officers could be requested to serve as raters. Two or three raters are sufficient. Instruct raters as to what to look for during the test and how to use the rating scales and forms. An average score should be calculated from all raters for each student. For instance, if a student received three scores of 97, 94, and 94 from three different scorers, the average score would be 95 (97 + 94 + 94 = 285 ÷ 3 = 95).

Name _____

Date _____

HOSE PRACTICALS

Evolution or operation: Hose load or finish designated _____

Loads for evolution: Horseshoe, reverse horseshoe, accordion, divided hose bed, skid load finish, donut roll finish.

Each load should be constructed with six (6) sections of hose within 3 minutes and 30 seconds for each load.

Items to be checked: In team operations, designate person making mistake. ( ) = points possible	Correct	Incorrect	Comments: Include name of person making mistake. ( ) = points scored

1. Starts hose load properly
   a. Reverse
   b. Forward
          (30)
			( )

2. Makes load or finish designated
   a. Correct procedure (15)
   b. Makes dutchman as required (15)
   c. Starts tier correctly (15)
   d. Staggers ends of folds (15)
			( )
			
			( )
			( )
			( )

3. Attaches nozzle(s)/adapters to load as finish requires (10)			( )
			( ) = total points scored

Time: 3 minutes and 30 seconds for each load.
Deduct one point for each second over suggested time.

SCORE _____

Figure 8.4 Example of a manipulative-performance test rating sheet.

Step 6: Follow established procedures during the administration of a performance test. The instructor should have all necessary apparatus and equipment ready before beginning the test. The instructor should also use the same equipment throughout the test, follow the same sequence of jobs for all students taking the test, and rate each student on the same basis. This will enable the instructor to concentrate on students rather than on duties of test administration. Nothing should be allowed to distract the instructor from carefully observing the student or the student's performance.

Step 7: Make a score distribution (preferably in percentiles) after tests have been administered and carefully evaluate students with low scores. Students who have difficulty in performing manipulative skills should receive immediate attention.

Step 8: In a team evaluation, it is critical to rotate team members in every position for rating. This way each student will be tested at each position in the evolution.

Advantages Of Manipulative-Performance Tests

Validity. Manipulative-performance test is the only valid method of measuring student achievement in learning manipulative skills.

Reliability. A properly constructed manipulative-performance test is a reliable measure of performance, if an appropriate rating scale and criteria are used.

Observation. Manipulative-performance tests permit observation of individual differences in judgment and approach to problems. Certain students may not be able to express themselves as well as others, either orally or in writing. However, they may be able to actually perform a job as well as, or better than, more verbal students.

Student Motivation. Manipulative-performance tests are excellent as a means of motivating students. Students who know they will be subjected to a manipulative-performance test will usually devote extra, outside-of-class time to prepare.

Sense of Accomplishment. Students who successfully complete a well-prepared and well-administered performance test can be proud of their accomplishment.

Disadvantages Of Manipulative-Performance Tests

Unreliability. Manipulative-performance test scores may be unreliable because of subjectivity in administering and scoring; that is, the reliability is proportionate to the quality of the identification of the criteria by the instructor. Lack of definitive rating criteria may add subjectivity and lessen reliability.

Economy of Test Taking. Manipulative-performance tests are time consuming. It may also take more instructors to monitor or evaluate the students.

Difficulty. It is harder to test groups in team evolutions.

Classified According To Purpose
PRESCRIPTIVE
Prescriptive tests are typically called pre-tests and are given at the beginning of instruction. They measure readiness or determine placement. When *measuring readiness*, the test should

answer the question, "Does the student have the skills needed to perform in the course?" If not, remedial activities should be offered or the student should accumulate more job experience.

When *determining placement*, the test should answer the question, "Has the student already achieved the behavioral objectives?" If so, the student should move on to the next course or unit of instruction. If not, the student can attend the class.

Prescriptive tests that measure readiness include each prerequisite skill required for entry. The test items are usually easy and criterion-referenced. Prescriptive tests that determine placement of students provide a representative sample of test items. The items generally have a wide range of difficulty and are norm-referenced.

PROGRESS

Progress tests are often viewed as quizzes, pop tests, or a question/answer period in class and are given throughout the course or unit of instruction. These tests typically *measure improvement* and give the instructor and students feedback on learning progress. When measuring improvement, the test should answer the question, "Is the student achieving the objectives?" Tests include the most important behavioral objectives or all of them, if possible. Each test item should match the level of difficulty of the corresponding behavioral objective, and thus will be criterion-referenced.

If learning problems are encountered during the course, the instructor may design a specific test to *diagnose learning difficulty*. When trying to diagnose causes of the problem, the test should answer the question, "If students are not achieving the objectives, what are the specific causes?" Test items represent a sample of tasks in which students are experiencing difficulty. The items are easy so that specific causes of difficulty can be pinpointed.

COMPREHENSIVE

Comprehensive tests are typically given in the middle or at the end of instruction. They *measure terminal performance*. Performance can be in the cognitive, affective, or psychomotor domain of learning. When determining terminal performance, the test should answer the question, "Have the students achieved the course objectives?"

A performance test may be used to assign grades, which makes it norm-referenced, or to certify mastery, which makes it criterion-referenced. When norm-referenced testing is the case, test items include a representative sample of course objectives and have a wide range of difficulty. This is not typical in the fire service, because professional qualification standards provide the basis for criterion-referenced tests.

Classified According To Administration Method

Oral tests, manipulative-performance tests, and written tests are grouped together based upon the various ways tests can be administered to students.

ORAL

During oral tests, the student generally gives verbal answers to spoken questions. Oral tests are not commonly used in the fire service but under certain circumstances the instructor may find them necessary. An example of this would be a student who cannot read at the level the test is written. Oral tests may also be used to supplement manipulative-performance tests to determine whether a student knows the reasoning behind the jobs performed. Oral tests are usually given one-on-one with the instructor.

When giving oral tests, the instructor must listen carefully to the student's responses since people will phrase answers differently. Oral tests should be used only when absolutely necessary and should not be used as a sole means of evaluating students or for evaluating terminal performance.

WRITTEN

Written tests may be subjective or objective and are useful in measuring retention and understanding of technical information. Written tests evaluate the accomplishment of cognitive and affective objectives. Fire chemistry, laws and ordinances, and hydraulic principles are examples of technical subjects for which students should be given written tests (Figure 8.5).

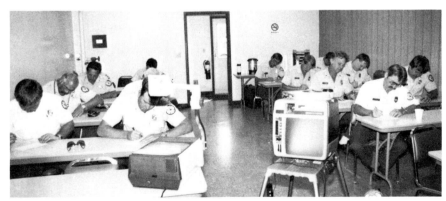

Figure 8.5 Written tests are useful in measuring retention and understanding of technical information.

Objective

Objective test items should be thoughtfully constructed. Often, the instructor limits objective test items to the lower cognitive learning levels of recall and recognition because these levels lend themselves easily to the objective format. However, objective test items can also be used to measure higher levels of cognitive learning such as interpretation and analysis. For example, an objective that asks the student to choose the correct function for a tool from among four listed functions requires only that the student recognize the correct function. On the other hand, a test item that presents a problem and asks the student to choose the best tool for solving the problem requires the student to interpret and analyze.

Subjective

A subjective test is one in which there is no set response. "Correct" responses vary with each student's solution to the problem. Because they allow the student the freedom to organize, analyze, revise, redesign, or evaluate a problem, subjective tests are effective for measuring the higher cognitive levels of analysis, evaluation, and interpretation. In the fire service, subjective tests — usually in an essay format — are used most often in officer training. At this level, most of the training involves intellectual, problem-solving skills.

TEST PLANNING

The key to designing effective tests is planning. Planning ensures that the instructor will measure the appropriate outcomes, include an adequate sample of the intended outcomes, and have the type of evidence needed to make sound instructional decisions. Test planning includes:

- Identifying and defining the behavioral objectives or learning outcomes

- Determining the purpose and type of the test

- Preparing the test specifications

- Constructing test items

If the first three steps are carefully done, constructing test items is simplified.

Identifying And Defining Behavioral Objectives

Course planning includes establishing objectives for the course that provide the instructor with a target. Objectives establish the desired results of a particular course. They also indicate testing and evaluation requirements. For example, if a course is designed to develop the ability to perform a series of jobs, the instructor must find out whether students who have completed the course can actually perform these jobs.

Behavioral objectives and the level to which each student must be trained are indications of what is to be measured. All planning should be directed toward what students will be able to accomplish as a result of instruction.

Determining The Purpose And Type Of The Test

The primary purpose of testing is to discover the amount of learning that has occurred due to instruction. Therefore, the test should match the behavioral objectives. If the objectives are directed toward a psychomotor skill, the test should cover that skill, not information about the skill.

However, there are other considerations to be made when determining the purpose of a test. One question to be asked is: Is the test to assess student performance against a set criteria or to rank performance against other students? The answer to that question will determine whether a criterion-referenced or norm-referenced test should be used.

Still other considerations are relevant to planning the type of test to be used. The answers to the following questions will help decide what type or types of tests should be used. Is the test to determine readiness for instruction or placement in the appropriate instructional level (prescriptive tests)? Is the test to measure improved progress or to identify learning problems that are hampering progress (progress tests)? Finally, is the test to rate terminal performance (performance tests)?

Another set of concerns is how an instructor plans to administer a test. Will the test measure manipulative skills (manipulative-performance tests) or technical skills (written tests)?

Preparing The Test Specifications

Although test specifications vary widely in their application and use, they are nonetheless a critical tool for the instructor. The most widely used approach is the test planning sheet. A well-prepared test planning sheet will assist the instructor in writing appropriate test items. The function of the test planning sheet is to specify the levels of learning in each content area. In norm-referenced tests, the specifications also guide the sampling of a sufficient number of test items. In criterion-referenced tests, one test item for each objective may be enough.

A test planning sheet is developed by: 1) identifying the levels of learning and the content material to be tested, 2) deciding on the relative importance of the content material at each level of learning (only for a norm-referenced test), and 3) making a two-way table to distribute the number of test items in each appropriate cell. The test planning sheet indicates the number and nature of the test items, therefore making test item writing much easier (Figure 8.6).

Constructing Test Items

Now comes the challenge of matching the specifications with individual test items. Some general considerations must be

(SAMPLE)
Test Planning Sheet

Course __Breathing Apparatus__   Purpose of Test __Quiz__

Type of Test __Written__   Total No. of Items __30__

Job or Topic	Instruction or Training Level	Testing Levels and Number of Items		
		Remembering (1)	Understanding (2)	Performing (3)
History of Breathing Apparatus	1	2		
Respiratory Hazards	2	1	2	
Principles of Operation	2	2	2	
Major Components, Their Purpose and Use	2	3	3	
Inspection and Servicing	3	1	2	4
Donning Breathing Apparatus	3			3
Working with Breathing Apparatus	3		2	3

Figure 8.6 A valuable aid for test construction is a test planning sheet.

looked at before detailed procedures are discussed in a later section for writing test items. The following are tips to help the instructor write strong test items.

Match items to course and behavioral objectives. Draft test items so they provide a test that will measure the intended learning outcome found in the lesson objectives.

Eliminate all barriers. Compose test items that eliminate or at least minimize barriers to taking the test. Some barriers to test taking include:

- A higher reading level than the student audience possesses
- Lengthy, complex, or unclear sentences
- Vague directions
- Unclear graphic materials

Avoid giving clues to test answers. Write items that avoid giving clues on how to answer the item correctly. Some areas to avoid include:

- Utilizing word associations that give away the answer
- Using "a" or "an," or plural or singular verbs that hint at the answer or eliminate the wrong answer
- Using words that make some answers more likely (for instance, sometimes) or others less likely (for instance, always or never)
- Placing the answer consistently in the same location (such as in the choice B answer in multiple-choice) or making the correct answers consistently longer (such as all true statements longer than false statements)
- Using copied material from the textbook or stereotypical answers
- Using test items that give the answer away in other items

Select the proper item difficulty. Since a function of criterion-referenced testing is to describe the precise learning outcome of a given student, item difficulty should match the course objectives. Norm-referenced testing, on the other hand, is designed to rank students against others in the class. Therefore, test items should range in difficulty to allow maximum discrimination among students.

Decide the number of items. Although the test planning sheet should tell how many test items should be used, there are practical constraints and educational considerations. Educationally, it is recommended for criterion-referenced tests that at least 10 items per learning level be used. This ensures that each level of learning is adequately measured. However, this is not to be confused with one test item per learning objective.

Practically, the number of items must match the time allowed for testing. As a rule of thumb, an adult can average one multiple-choice, three short answer/completion, or three true-false items in one minute. Remember that the slower students in the class will take more time. Use experience as the best guide to the number of items for the time allotted.

The number of items is the key to comprehensiveness in test writing. A test should be constructed so it can measure the student's ability in all phases of the course. It should be complete without unnecessary details.

Check for ease. Design tests that are easy to give, easy to take, and easy to score. To be easy to give, a test should be constructed so the instructor can concentrate on students during the testing period. A test that burdens the instructor with the mechanics of administration is an obstacle to effective testing.

To be easy to take, a test must include directions that are clear and complete so class members will know exactly what they are to do. Sample test items should be provided to show the class how the test is to be answered.

To be easy to score, a test must be well constructed and specific, with no ambiguities in the items or the directions. An answer sheet or simple test form will help with the task of scoring.

Build in validity and reliability. The two most important conditions of a well-designed test are validity and reliability. Validity is the extent that a test measures what it says it is to measure. It is built into the test by selecting an ample number of test items for each learning level and content area. The best way to ensure validity of the course content is to 1) identify the content of the course and the behavioral objectives to be measured, 2) develop a table of specifications, which specifies the sample of test items to be used, and 3) design a test that matches the specifications.

Reliability is the consistency and accuracy of measurement in a test. A reliable test is free of ambiguous items or directions, vague scoring criteria, environmental distractions, and opportunities for cheating or guessing. A reliable test is one in which separate scorers would give the same score to the same student's test. It also is one in which, if a student were to take the test one day and take the same test three days later, the student would make the same score. Devoting attention to each of the foregoing test characteristics, analyzing a test each time it is given, and discarding or rewriting items that do not meet requirements should aid in improving test reliability.

Any test that cannot give consistent and accurate scores cannot be measuring what it said it would measure. Therefore, reliability is an essential condition of validity, but is not the only condition of validity.

TEST ITEMS
Multiple-Choice

A well-constructed multiple-choice test item is generally recognized as one of the most versatile of the objective test items. It can measure a variety of the student's abilities and can be adapted to most types of subject matter.

Multiple-choice test items consist of either a question or an incomplete statement, commonly referred to as the *stem* of the item. Following the stem is a list of several possible answers, referred to as the *choices* or *alternatives*. The student is asked to read the stem and to select the correct answer from the list of choices. The correct choice in each item is known as the *answer*, and the remaining choices are called *distracters*. The obvious purpose of distracters is to distract those students who are not sure of the correct answer (Figure 8.7 on the next page).

(SAMPLE)
TEST FORMAT
Multiple Choice

Name _____

Date _____

DRIVERS TRAINING
Test

INSTRUCTIONS: This is a multiple-choice test. For each of the following questions or statements, draw a circle around the letter preceding the *one* correct answer.

EXAMPLE: Safe driving skill is:
  A. hereditary
  B. a matter of training
  C. a responsibility of administration
  D. an attitude

1. The average reaction time in seconds on the highway is:
  A. 0.1
  B. 0.14
  C. 0.44
  D. 0.75

2. The psychophysical test helps in the selection of drivers. Which one of the following is not a measurement of this test:
  A. peripheral vision
  B. depth perception
  C. glare acuity
  D. color vision

3. Which one of the following is true of Braking Distance?
  A. it is roughly proportional to the square of the driving speed
  B. it does *not* change regardless of speed
  C. it decreases with speed
  D. it increases at a slower rate than the speed of the vehicle

Figure 8.7 Example of a multiple-choice test.

Advantages of Multiple-Choice Test Items

Widely Applicable. These test items are one of the most widely applicable test items for measuring achievement.

Flexible. These test items can measure complex learning outcomes as well as various types of knowledge.

Avoidance of Ambiguity and Vagueness. The ambiguity and vagueness that are common in other types of test items are avoided.

Ease of Construction. In general, multiple-choice items make it easier to construct high-quality test items.

Disadvantages of Multiple-Choice Tests

Limitation of Measurement. These test items are not well suited for measuring some skills, such as the ability to organize and present ideas.

Difficulty in Constructing. It may be difficult to locate a sufficient number of incorrect but plausible answers.

Guessing. The student can guess a correct answer yet not know the material.

WHAT TO INCLUDE IN MULTIPLE-CHOICE TESTS

- An introductory statement (stem) in the form of a direct question or an incomplete statement to measure only one learning outcome

- A stem as clear and brief as possible

- As much of the wording in the stem as possible

- The wording of the stem in positive form, when possible

- Negative words underlined in the stem and in the choices

- Four choices: the answer and three plausible distracters

- An answer that is clearly best and cannot be argued

- Choices that all refer to the same subject matter

- Choices that are grammatically consistent with the stem of the item and parallel in form

- Plausible, yet attractive distracters

- Varied position of the answer

- Choices on separate lines and in a column

- Choices indented five spaces

- Capital letters for identifying the choices

- Double letter spacing between the identification letters and the choices

- Periods after the item numbers and the choice identification letters

- No periods after the choices

WHAT TO AVOID IN MULTIPLE-CHOICE TESTS

- Choices that are obviously wrong.

- Grammatical errors that give clues to the correct answer (examples are the use of *a* and *an* and of mixed singulars and plurals).

- Correct answers that are consistently longer or shorter than the distracters or uniformly placed (such as all correct answers being the choice of "b").

- All of the above as a fourth choice; use none of the above with care.

True-False

Probably the best known of the various types of objective test items is the true-false, or alternative-response, item. Although considered relatively easy to construct, true-false items are also the most abused, and their quality is often doubtful.

In addition to the traditional true-false items (Figure 8.8), there are also the modified true-false items. The modified item asks the student to explain why an item is false, if that is the answer given (Figure 8.9). These items test learning at higher levels.

(SAMPLE)
TEST FORMAT
True -False

Name _____

Date _____

CHEMISTRY AND PHYSICS OF FIRE
Test

INSTRUCTIONS: This is a true-false test. If the statement is true, draw a circle around the "T"; if false, draw a circle around the "F".

EXAMPLE: (T) F   Fire is a chemical reaction.

T F 1. The molecular weight of a compound refers to all the atoms in a molecule of that specific compound.

T F 2. When a combustible solid catches fire it has reached its ignition temperature.

T F 3. Atmospheric pressure at sea level is 14.7 psi.

T F 4. Mineral oils will oxidize at ordinary room temperature.

T F 5. Solids will NOT change to a vapor at ordinary room temperature.

Figure 8.8 Example of a traditional true-false test.

The true-false item usually consists of a single statement that the student is required to recognize as either true or false. However, difficulty is usually encountered in constructing items that are completely true or completely false.

Advantages of True-False Test Items

Adaptability. True-false test items are well suited to sample a wide range of subject matter.

(SAMPLE)
TEST FORMAT
Modified True -False

Name _____

Date _____

CHEMISTRY AND PHYSICS OF FIRE
Test

INSTRUCTIONS: This is a true-false test. If the statement is true, draw a circle around the "T"; if false, draw a circle around the "F". If the statement is false, explain *why* it is false in the space below it.

EXAMPLE: T   Ⓕ   Fire is a physical reaction.
                    <u>Fire is a chemical reaction.</u>
                    _____

T   F   1. The molecular weight of a compound refers to all the atoms in a molecule of that specific compound.
          _____
          _____

T   F   2. When a combustible solid catches fire it has reached its ignition temperature.
          _____
          _____

T   F   3. Atmospheric pressure at sea level is 14.7 psi.
          _____
          _____

Figure 8.9 Example of a modified true-false test.

Economy of Test Taking. True-false test items are brief enough to permit a large number of items to be answered in a short time.

Ease in Scoring. True-false test items are easily scored.

Student Motivation. If constructed with care, true-false test items can serve to promote interest and to introduce points for discussion.

Disadvantages of True-False Test Items

Difficulty in Constructing. The ease with which true-false test items are constructed is frequently cited as an advantage. This more than likely has resulted from the simple process of taking statements from textbooks, altering half of them to produce false statements, and presenting the final product to the students as true-false test items. In reality, however, it is easy to construct poor true-false items. Constructing good true-false items is a difficult task.

Guessing. A student has a 50/50 chance of being able to answer items correctly whether the student knows anything about the subject matter.

Ambiguity. It is difficult to avoid ambiguities and to construct items that are either completely true or completely false without making the correct answer obvious. This tends to limit true-false test items to factual details only.

Limitation of Expression. Students sometimes express the belief that true-false test items have not given them the opportunity to demonstrate what they really know or can do.

WHAT TO INCLUDE IN TRUE-FALSE TESTS

- The letters T and F at the left margin, if answers are to be marked on the test paper. Direct students to draw a circle around the answer they select. Do not direct students to write in T or F, + or -, or yes or no.

- A sufficient number of items to provide reliable results.

- Equal number of true and false items distributed randomly throughout the test.

WHAT TO AVOID IN TRUE-FALSE TESTS

- Specific determiners. *Usually, generally, often*, or *sometimes* are most likely to appear in true statements. *Never, all, always*. or *none* are more likely to be found in false statements. Specific determiners provide unwarranted clues.

- Trick or catch items.

- Double-negative statements.

- The personal pronoun you, and command statement.

- Trivial statements. The instructor should strive to develop items that require students to think about what they have learned, rather than to merely remember it.

- True items that are consistently longer or shorter than false items.

- Long, complex statements.

- Humorous and absurd statements.

- Statements that are taken directly from textbooks.

Matching

Matching test items, in a sense, are a variation of multiple-choice test items. They are relatively easy to construct, can be objective, and are easy to score. They are especially applicable for testing for the who, what, where, and when types of subject matter.

Matching test items consist of two parallel columns of words or phrases. The student is required to match each item in one column with the response in the other column to which it is most closely related. Some examples of matching test items are shown in Figures 8.10 and 8.11 on next page.

A partial list of items that can be related in matching test items includes the following (some of these can be pictorial):

- Short questions — with their answers
- Parts — with their functions
- Terms — with their definitions
- Objects — with their names
- Machines or tools — with their uses
- Problems — with their solutions
- Causes — with their effects

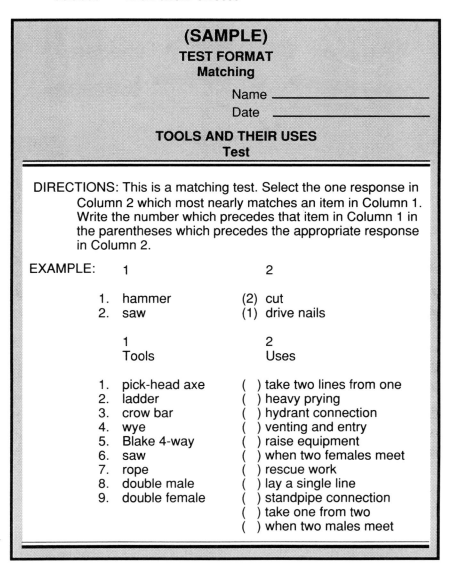

(SAMPLE)
TEST FORMAT
Matching

Name _____

Date _____

TOOLS AND THEIR USES
Test

DIRECTIONS: This is a matching test. Select the one response in Column 2 which most nearly matches an item in Column 1. Write the number which precedes that item in Column 1 in the parentheses which precedes the appropriate response in Column 2.

EXAMPLE: 1   2

1. hammer   (2) cut
2. saw   (1) drive nails

1   2
Tools   Uses

1. pick-head axe   ( ) take two lines from one
2. ladder   ( ) heavy prying
3. crow bar   ( ) hydrant connection
4. wye   ( ) venting and entry
5. Blake 4-way   ( ) raise equipment
6. saw   ( ) when two females meet
7. rope   ( ) rescue work
8. double male   ( ) lay a single line
9. double female   ( ) standpipe connection
                   ( ) take one from two
                   ( ) when two males meet

Figure 8.10 Example of a matching test.

(SAMPLE)
TEST FORMAT
Pictorial Method

Name ____________________
Date ____________________

TOOL IDENTIFICATION
Test

DIRECTIONS: This is a matching test. Select the one response in Part 2 which most nearly describes the tool in Part 1 and place the appropriate letter from Part 1 in the parentheses preceding the response in Part 2.

Part 1

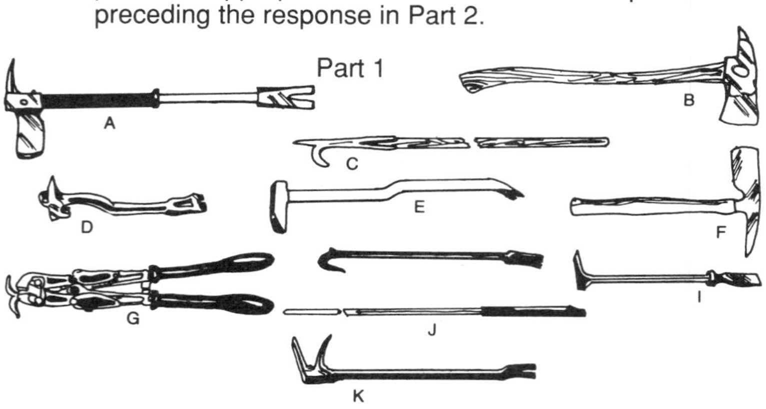

Part 2

( ) Pick-head axe
( ) Halligan
( ) Pike pole
( ) Pry-Axe
( ) Kelly tool
( ) Spanner wrench
( ) Claw tool
( ) Wrecking bar
( ) Wire cutter
( ) Hatchet

Figure 8.11 Example of a matching test.

Advantages of Matching Test Items

Elimination of Guessing. When matching test items are properly constructed, the guessing factor is practically eliminated.

Economy. Matching test items are relatively easy to construct and easy to score. In addition, they can cover a large amount of factual material in a compact space.

Disadvantages of Matching Test Items

Limitation of Measurement. Because the phrases used must be short, matching tests provide a poor measure of complete understanding. They are considered inferior to multiple-choice tests in measuring higher levels of instruction.

WHAT TO INCLUDE IN MATCHING TESTS
- All matching responses on one page.
- Test arranged in two columns (Column I should be at the left side of the page, Column II at the right side of the page).
- The numbered items in Column I with space for an answer before the number — the lettered responses in Column II.
- Directions stating whether a response may be used more than once.
- At least 5, but not more than 12, items in Column I.
- Two or three more responses in Column II than items in Column I, so that it requires more precise knowledge from the student answering the item.
- Specific instructions.

WHAT TO AVOID IN MATCHING TESTS
- Choices that are obviously wrong.
- Column I and Column II on more than one page.
- The same number of items in Column I and Column II.

Short Answer/Completion

The short answer/completion test item is the only objective test item that requires the student to supply the answer. It is constructed similarly to the multiple-choice test item, but without the alternatives from which to select. The short answer item is a question to which an answer is required. The completion item is an incomplete statement with key words omitted. The student is to recall previous learning, apply principles, understand methods or procedures, and insert the answer (Figure 8.12 on next page).

Advantages of Short Answer / Completion Tests

Flexibility. These test items are commonly used in measuring who, what, when, where, and how information. They are also useful in measuring a student's ability to solve mathematical and similar problems.

Elimination of Guessing. One of the most important advantages of short answer/completion test items is that the student must supply the answer. This minimizes the possibility that the student will be able to guess the correct answer. Partial knowledge, which might enable the student to choose the correct answer from a list of choices, is usually insufficient to answer a short answer/completion item.

(SAMPLE)
TEST FORMAT
Short Answer

Name _____
Date _____

FIRST AID
Test

DIRECTIONS: This is a short answer test. For each of the following items, enter the correct answer in the numbered space provided.

EXAMPLE: How many times per minute does a normal adult's heart beat?   ____72____

1. What is the term commonly used which means that death is beginning in the brain?   1. _____

2. What is the term used to denote the cooling process of the body following death?   2. _____

3. Rigor mortis, a tightening of the muscles following death, usually begins in what muscles of the body?   3. _____

4. What is the name of the nerve center in the brain that controls breathing?   4. _____

5. During normal breathing, how many pints of air are exchanged with each complete respiration?   5. _____

Figure 8.12 Example of a short answer test.

Disadvantages of Short Answer/Completion Tests

Limitation of Measurement. These test items cannot measure complex achievement.

Difficulty in Constructing. It is hard to design the item so that only one answer is right, and it is clear to the student which term is to be used for the answer. For example, if asked, "What is the primary responsibility of the company officer?" which answer could the student use — supervise, manage, lead, or delegate? It must be clear to the student which answer is correct or the student should receive credit for any answer that means the same thing.

Difficulty in Scoring. A variety of answers may appear and must be considered for total or partial credit. In addition, spelling problems may also arise. Misspelled words may make it difficult to determine whether the answer is correct.

WHAT TO INCLUDE IN SHORT ANSWER/COMPLETION TESTS

- Short, direct items so that only one answer is possible. Start with a direct question and change to an incomplete statement when it appears to be more concise. Make sure the answer to be supplied is a key point in the lesson.

- A blank at the end of the sentence.

- Instructions on the degree of precision and the units to be used to express the answer.

WHAT TO AVOID IN SHORT ANSWER/COMPLETION TESTS

- Long, involved statements

- Answers that call for more than one word, number, or phrase

- Statements copied from the textbook

- Any unnecessary clues, like variations in the length of the blanks for answers, or the use of "a" or "an"

COMMON CONSIDERATIONS FOR ALL TESTS

Format

For all tests, there are certain format considerations that make test administration, test taking, and test scoring simpler. The following should be included in any test format:

- Spaces for the date and the student's name (these are not necessary on the test sheet if a separate answer sheet is to be used)

- Test title or label

- Each test handed out numbered for reporting scores and for test security

- Clear, easy-to-follow instructions at the beginning of the test

- A sample item to demonstrate how the test is to be taken

- Consecutive numbering of items

- Single spacing between lines within items

- Double spacing between items

Arrangement

After the test items are prepared, they must be arranged in a logical way. They can either be grouped by learning outcome (such as basic, intermediate, or advanced) or by type of test item (such as multiple-choice, matching, or short answer). In either case, it is recommended that the items be placed in sequence of increasing difficulty. This allows the student to answer some easy items, gain confidence, move on to the moderately difficult items, and then be challenged by a few more difficult items toward the end of the test.

Instructions

Instructions for taking a test should be brief and to the point. They should explain the purpose of the test to the student. They should also explain how the student is to record answers and whether to guess when in doubt of the answer. They should also state the time allowed to complete the test. Often, it is good policy to read the instructions aloud as the class reads along, then ask if there are any questions. Where questions are grouped by the type of test item, it is frequently desirable to give specific instructions at the beginning of each section.

Security

Cheating on tests presents special problems for instructors, problems that may be reduced by careful attention to a few details. The instructor can protect test security by regularly revising tests and by exercising care in typing, duplicating, and storing test materials. Most cheating is a result of advance knowledge of test content, caused by the instructor's inattention to these details. To improve test reliability and validity, the instructor must ensure that test scores are the result of the students' own unaided efforts.

Under certain conditions, it may be necessary to discourage cheating by creating special seating arrangements and by carefully supervising students during the test. A method used to ensure that test booklets are returned is to number them and ask that they be returned with answer sheets.

Administration

Administration of a test begins before the test is to be given. Students should be notified in an informative and motivating manner that a test will be given, when it will be given, what it will cover, how they might prepare, and what they should bring to class for the test (paper, pencil, pen, notes, or books).

On the day of the test, environmental conditions should be conducive to the students' best efforts so the quality of their performance is not influenced. To aid students in reaching their potential while taking a test, the instructor should arrange, when practical, for suitable physical conditions. These may include proper room temperature, adequate lighting, and proper seating in an area free from distractions or noises.

Instructors are aware that motivation affects student performance, although environmental factors are the usual concern for instructors at test time. The instructor needs to place emphasis on alleviating stress and anxiety, while providing a motivating atmosphere for taking tests. This can be achieved by displaying a pleasant demeanor, by not spending too much time in giving instructions and taking time away from test taking, by not hurrying students, or by not threatening students with a test as a disciplinary measure. Tests alone generate stress for the student. The instructor's job is to eliminate or at least keep stress to a minimum during test taking time.

The final step in administering a test is to report the results and decide how to use them for further learning. The results can be reported on a class-wide basis, so each student knows how he or she compares, if it is a norm-referenced test. Individual results can be shown in comparison to course objectives and the total number of points possible, if it is a criterion-referenced test. Tests should be reviewed so students can learn which test items they missed and why.

TEST ITEM ANALYSIS

A test item analysis is a helpful tool to the instructor. It can show how difficult a test is, how much it discriminates between high and low scorers, and whether the alternatives used for distracters work to distract students. This process makes the instructor aware of problem test items so they can be improved.

The simplest way to analyze a test item is to list each item with the possible answers. Then count and record the number of times each possible answer was selected. The correct answer is boldfaced in this example.

	A	B	C	D
1.	5	8	**6**	1
2.	2	3	2	**13**
3.	**2**	17	1	0

By using this example for question #1, an instructor can see that only six students chose the right answer, choice C. Since eight students selected the B distracter, it can be assumed that it served as an effective distracter. That is because those who were unsure of the answer chose a distracter that seemed plausible, but was not as strong as choice B. The A distracter worked with five students selecting that choice. The D distracter had limited effectiveness. However, if only six out of twenty students answered the item correctly, then the question may not have tested the intended learning.

In test item #2, the majority of the class chose D for the correct answer. This question did not distinguish among students very well, since there were few students who answered incorrectly. The instructor will have to use some judgment on this item. It cannot be discerned by this information whether the item was too easy or whether it accurately tested for learning.

The third test item shows the power of choice B to distract students from the correct answer. However, choices C and D are poor distracters. It is possible that they should be discarded and new distracters written.

Analyzing Test Results

A complete evaluation process must include behavioral objectives, test results, and a decision about whether the results measure the objectives. The primary purpose of analyzing test results is the same as the purpose of evaluation — to improve the teaching/learning process. Analyzing test results is the process of using different methods that help interpret the results for the instructor. These methods range from simple grading to ranking scores to describing the measures of central tendency and the variability of scores.

If a class scores poorly, there are some steps an instructor can take to correct the skewed test scores. One, the instructor can throw out poor test items. Two, the instructor can reteach the lesson and retest. Three, the instructor can review and adjust the test items by changing the distracters, then giving the test again. Another possibility is to review the instruction and the circumstances (such as a fire call occurring). Then adjust the instruction, reteach, and retest. If two classes are meeting at the same time, another option is to compare their test results. Then decide what steps need to be taken.

VALIDITY AND RELIABILITY

A secondary purpose for analyzing results of tests is to determine if validity and reliability are maintained. If the results are not valid or reliable, they have no meaning to the instructor. As can be recalled from previous sections of this chapter, validity is the degree to which the test measures what it sets out to measure. Reliability is the consistency of test scores from one measurement to another.

Reliability is a condition of validity. In other words, if test scores differ on a test given one day from a test given the next day covering the same material, then the test scores are not reliable. If the scores differ, they cannot be said to measure what they set out to measure and are not valid. Therefore, reliability is necessary to validity.

Criterion-Referenced vs. Norm-Referenced Tests

Validity and reliability have different meanings when interpreting the results of criterion-referenced and norm-referenced tests. In both types of tests, validity means the extent to which the test measures student achievement of the behavioral objectives. In criterion-referenced tests, validity refers to the measurement of mastery or nonmastery by an individual as compared against behavioral objectives. In norm-referenced tests, validity refers to the measurement or grade of the individual student as compared against other students in a class.

Reliability of criterion-referenced tests means the consistency of results to classify mastery or nonmastery by an individual. In contrast, reliability of norm-referenced tests means the consistency of results among students. So it is clear that the type of test, criterion-referenced or norm-referenced, should be specified and designed by intent, not by accident.

COMPOSITE SCORING

The basis of assigning scores or grades should be a composite of various course activities and other factors. Specific values should be established for assignments, projects, quizzes, examinations, and so forth. Other factors to consider may include attitudes, cooperation, participation, and other less obvious learning outcomes. These will normally come from the affective learning domain.

POINT SYSTEMS

A system that is more objective and works for scoring both criterion-referenced and norm-referenced tests is the point system. The instructor establishes point values for each course activity, then totals the points for the course score. Any number of points may be used, and the total points earned by a student can be converted to a letter grade in norm-referenced testing.

- 85-100 percent of possible points = A
- 75-84 percent of possible points = B
- 65-74 percent of possible points = C
- 50-64 percent of possible points = D

The point values can be converted to a mastery or nonmastery classification in criterion-referenced testing.

- 85-100 percent of possible points = Mastery
- 84 percent or less of possible points = Nonmastery

The instructor must be as objective as possible in scoring and grading, and must not be influenced by factors not pertinent to the student's achievement. The instructor should be prepared to defend any score or grade given. Developing a standard system of scoring and grading, then sticking to it will serve to increase objectivity, consistency, fairness and therefore, reliability. See the rating scale used to observe and rate student class participation in an objective way.

CLASS PARTICIPATION RATING SCALE

The class participation scale is used to document the active participation of students in a class. In the example below, the instructor would circle the number which best describes the student in class. The numbers represent the following values: 5 = excellent; 4 = good; 3 = acceptable; 2 = needs improvement; 1 = not acceptable.

1. To what degree does the student participate in class discussions?

$$1 \quad 2 \quad 3 \quad 4 \quad 5$$

2. To what degree are the student's comments related to the topic of discussion?

$$1 \quad 2 \quad 3 \quad 4 \quad 5$$

3. To what degree does the student pose thoughtful questions on the topic of discussion?

$$1 \quad 2 \quad 3 \quad 4 \quad 5$$

The judgment by the instructor is subjective and determined by consciously observing students during class sessions. Instructors should realize when assessing participation levels that some students are more aggressive than other students. This aggressiveness may make the docile students appear unusually intraverted. The instructor must compare and evaluate each class member as an individual while simultaneously comparing them to the other members of the class.

STATISTICAL METHODS OF ANALYZING TEST RESULTS

Methods of analyzing test results are generally referred to as statistics. Statistics are nothing more than a way of organizing, analyzing, and interpreting test scores. Elementary statistical methods described here use simple arithmetic skills the instructor already possesses. The only two new items are the introduction of new terms, which is to be expected in any new learning, and the use of statistical symbols, where symbols stand for words or brief descriptions. The instructor can begin with known factors and move into the unknown areas.

Raw scores, which are what the instructor starts with, are the points received on a test. If a student answers 38 items correctly out of 40 test items, then the raw score is 38. Another way of reporting test results is by a percentage score. A percentage score is computed by adding the number of correct answers, and dividing by the total possible. In this case, the *percentage score* is 95 percent, or 38 divided by 40.

The raw score is of little value in norm-referenced tests unless it is converted into some type of calculated score that shows how it compares to other scores from the same test in the same class.

Ranking of scores is a comparative approach, and there are several methods that can be used.

It is beyond the scope of this manual to teach a complete course in statistics; however, statistical analysis is valuable to instructors wanting to determine class ranking. It is recommended that instructors contact and enroll in an elementary statistics course at a local community or junior college. An elementary statistics course usually has a prerequisite of basic algebra. Most of the computations will be basic mathematics, that is, addition, subtraction, multiplication, and division.

EVALUATION OF COURSE AND INSTRUCTIONAL DESIGN

The primary purpose of evaluation in training is to improve the teaching/learning process. Often, instructors use only test results for evaluation. But there are many other types of evidence that collectively give a more complete picture as to whether the teaching/learning process was successful.

Two kinds of evaluation provide an approach for looking at the process and the product of the instructional process. Formative evaluation looks at the process, while summative evaluation looks at the product. In both cases, the evaluation process should be useful to its users. It should be easy and practical to administer and report. Data collected should be accurate.

Planning Course Evaluation

To get the most from a course evaluation, the instructor should plan the approach by thinking through the following questions:

- What questions need to be answered? How did participants feel about the training? What did they learn? How did training affect their attitudes and behavior? What were the organizational results?

- How can the items addressed above be answered? Will information gathering be administered by paper and pencil tests, questionnaires, or surveys? Will tests require participants to demonstrate their new knowledge and skills in a role play, simulation, or actual performance?

- What are the objectives of the training program? Are the evaluation criteria based on these objectives?

- Do the criteria indicate improvement between expected and actual performance when measured against the results of the needs analysis?

- What data sources are already available to help measure results (productivity reports, daily log sheets, training and personnel records)?

- Are there alternative methods for gathering this data such as interviews and on-site observations?

- What are the best and most cost-effective methods for measuring the results of the training? Are there less costly, more efficient ways of administering the evaluation?

There are six essential areas to a thorough evaluation of the instructional process:

REACTION — Were the participants satisfied with the course? Was management satisfied with the learning that occurred?

KNOWLEDGE — What new knowledge did the students acquire and demonstrate?

SKILLS — What new skills did the students acquire and demonstrate?

ATTITUDES — How has the training changed their opinions, values, and beliefs?

TRANSFER OF LEARNING — How has the training affected the way participants perform on the job?

RESULTS — How has the training contributed to accomplishing organizational goals and objectives?

In evaluating the instructional process, it is critical to make clear to anyone involved *what* is to be evaluated and *why*. This clarifies and guides how evaluation is to be conducted and eliminates misunderstanding.

Instructional design is the analysis of training needs, the systematic design of teaching/learning activities, and the assessment of the teaching/learning process. As training needs change, the instructional process will change. When the design of teaching/learning activities is no longer effective, the process must change. This is the reason for continuous assessment of the teaching/learning process. The seasoned instructor will always be alert to ways to improve or update instruction, or even eliminate instruction that is no longer needed. In a changing world of instructional needs, content, methods, and techniques, the instructor must remain flexible.

Formative Evaluation

Formative evaluation is the ongoing, repeated checking during course development and during or after instruction to determine the most effective instructional content, methods, aids, and testing techniques. Formative evaluation answers the question, "Am I teaching the right content and using the most appropriate methods to facilitate learning?" Or "Have the students learned in the most efficient way possible?"

FIELD TESTING

The course objectives still serve as the primary criteria or standards against which a judgment is made in light of the various types of evidence. Whether or not a course has been well designed to teach the course, objectives can be determined in a field test before the course has been finalized. A field test is merely teaching the course on a trial basis to see if the sequence of material facilitates learning, if the teaching methods and aids are appropriate and efficient in teaching the material, and if the testing procedures are adequate and appropriate for the course objectives. The course then can be revised before using it on a regular basis.

OBSERVATION

Evaluation is not a static, one-time event; it should occur at all times throughout instruction. Just as testing occurs throughout instruction, course evaluation occurs before, during, and after instruction.

The instructor can be gathering different kinds of evidence during instruction, most of which is observational. The following are factors the instructor should be attentive to:

- Student interest in subject
- Level of general participation
- Reactions to exercises and activities
- Level of student questions and comments
- Level of student frustration
- Level of student sense of achievement
- Results from quizzes

With any feedback during instruction, the instructor can change or modify instruction to meet the needs of a particular class. In some cases, it will be clear that the course needs to be changed on a permanent basis. The judgment of the instructor is critical to decisions about instruction.

Summative Evaluation

Summative evaluation is a one-time (at the end of the course) appraisal of the learning that has occurred. This usually is conducted in the form of the testing technique but can also include other techniques. Examples might include: observation of unexpected learning outcomes or follow-up survey to students or supervisors of retained learning. Summative evaluation answers the question, "Have the students learned what is needed to conduct their work back on the job?" Or "Has instruction been effective to meet the behavioral objectives?"

Evidence is gathered from two major sources: the test results and the course critique. The test results measure the degree of learning that has occurred. The course critique describes the student's opinion of the success of the course materials and the instruction. Sometimes instructor observations are taken into account in a summative evaluation.

COURSE CRITIQUE

A final course critique is an important part of the entire evaluation process to determine whether course objectives have been met. As a course of study or period of training ends, most instructors feel somewhat unsure of how well they taught and how much their students learned. This condition may always exist, but it can be reduced to some degree by an in-depth critique of the course. In addition to helping the instructors determine how well they and their students met the objectives, a final critique can aid in increasing student involvement and in improving course materials, teaching techniques, and training aids (Figures 8.13 through 8.16).

USING THE RESULTS

Unless the results collected from tests, course critiques, and observations are used to make some decisions about future instruction, then time, money, and effort have been wasted. The evaluation process should result in the following three steps:

Step 1: Determine causes for student failure.

- Inappropriate course objectives.
- Course outline and materials did not match objectives.
- Instructional methods and aids did not facilitate learning.
- Presentation style hampered learning.
- Learning environment stifled learning.
- Evaluation techniques were not valid or reliable.
- Personal or outside problems.
- Learning problems.

Step 2: Identify actions to be taken to correct deficiencies.

- Revise the course objectives.
- Modify the course outline and materials to match objectives.
- Change instructional methods/aids to facilitate learning.
- Improve presentation style to foster learning.
- Enhance the learning environment to promote learning.

COURSE EVALUATION QUESTIONAIRE

Title of course you are attending: _____

Marking: Circle choice as follows:

SA - If you *strongly agree* with the item.
A - If you *agree moderately* with the item.
D - If you *disagree moderately* with the item.
SD - If you *strongly disagree* with the item.

PLEASE READ EACH ITEM CAREFULLY

1. SA A   D SD   I would take another course taught this way.
2. SA A   D SD   Not much was gained by taking this course.
3. SA A   D SD   The course encouraged development of new viewpoints and appreciations.
4. SA A   D SD   The course material seemed worthwhile.
5. SA A   D SD   The instructor(s) demonstrated a thorough knowledge of the subject matter.
6. SA A   D SD   The course material was too difficult.
7. SA A   D SD   This was one of my poorest courses.
8. SA A   D SD   The course content was excellent.
9. SA A   D SD   Overall the course was good.
10. SA A   D SD   I think appropriate members of my department should attend this course.

Is this your first Service Institute course?   Yes _____ No _____

Are you a (check one)paid ___,paid-on-call ___ , volunteer firefighter ___?

How long have you been a firefighter?   _____ years.

What is your rank or title? _____

Figure 8.13 Example of a course evaluation questionnaire.

COURSE EVALUATION

1. Was this course what you expected it to be?   If no, why not?

   _____
   _____
   _____

2. What areas of instruction do you think could be added to the course to improve it?

   _____
   _____
   _____

3. What areas do you think could be shortened?   Entirely eliminated?

   _____
   _____
   _____

4. Were the exercises appropriate for the course?   If no, why not?

   _____
   _____
   _____

5. What did you particularly *dislike* about the course?

   _____
   _____
   _____

6. What did you particularly *like* about the course?

   _____
   _____
   _____

Figure 8.14 Example of a course evaluation questionnaire.

STUDENT EVALUATION OF INSTRUCTION

Instructor's Name _____ Course _____ Date _____

DIRECTIONS: Read the entire sheet BEFORE you mark any response. In the left-hand margin number the items 1, 2, and 3 that you consider to be the three most significant on the sheet. Place a cross (x) at the point on the scale that most accurately represents your considered opinion of *each* trait. DO NOT sign your name to this paper. Your fair and honest opinion is what really counts, as your instructor desires this rating for their own self-improvement.

KNOWLEDGE OF THE SUBJECT
( )   ( )   ( )   ( )
Very well informed   Well informed   Limited background   Poorly informed

PRESENTATION
( )   ( )   ( )   ( )
Stimulating   Adequate   Routine   Dull

ATTITUDE TOWARD STUDENT
( )   ( )   ( )   ( )
Very considerate   Considerate   Sometimes intolerant   Inconsiderate and rude

EXPLANATIONS
( )   ( )   ( )   ( )
Very clear   Clear   Confused   Faulty

POISE
( )   ( )   ( )   ( )
Highly poised   Poised   Easily upset   Highly insecure

ORGANIZATION OF COURSE
( )   ( )   ( )   ( )
Well organized   Organized   Lacks continuity   Confused

ASSIGNMENTS
( )   ( )   ( )   ( )
Very clear   Clear   Indefinite   Very vague

EXAMINATION QUESTIONS
( )   ( )   ( )   ( )
Clear and relevant   Adequate   Sometimes confusing   Irrelevant and not clear

GRADING METHODS
( )   ( )   ( )   ( )
Very fair   Fair   Inconsistent   Biased

TIME STUDENT SPENDS ON COURSE
( )   ( )   ( )   ( )
More than any other   More than average   Less than average   Less than any other

TEXTBOOK VALUE
( )   ( )   ( )   ( )
Great   Some   Limited   Very little

ATTITUDE TOWARDS COURSE
( )   ( )   ( )   ( )
Very favorable   Favorable   Indifferent   Negative

Figure 8.15 Example of an instructor rating sheet.

INSTRUCTOR RATING SHEET

INSTRUCTIONS: Write the name of the instructor(s) to be rated in the blank preceding the rating scale. Make a check in the appropriate boxes for each subject matter and teaching. When checking subject matter ask yourself, "will the material be of value to me and my department now or in the future?" Your check concerning teaching indicates how well you felt the instructor presented the material. Remember, material may be well presented but useless or very useful and badly presented. The following scale is for your use in checking.

Subject Matter	Teaching
1 = of no use	very bad
2 = of limited use	inadequate
3 = of some use	adequate
4 = of considerable use	good
5 = very useful	excellent

_____
(instructor)

	1	2	3	4	5
Subject Matter					
Teaching					

Comments: _____

_____
(instructor)

	1	2	3	4	5
Subject Matter					
Teaching					

Comments: _____

_____
(instructor)

	1	2	3	4	5
Subject Matter					
Teaching					

Comments: _____

_____
(instructor)

	1	2	3	4	5
Subject Matter					
Teaching					

Comments: _____

_____
(instructor)

	1	2	3	4	5
Subject Matter					
Teaching					

Comments: _____

Figure 8.16 Example of an instructor rating sheet.

- Alter evaluation techniques to produce valid and reliable instruments.
- Support student personally and offer referral services.
- Provide extra learning opportunities such as learning activity packets, individual or pair assignments, group study, or one-on-one tutoring.

Step 3: Document and report the findings to superiors.

- Keep daily records of training activity.
- Maintain individual training records.
- Retain class test results and analysis of results.
- Provide written progress reports to management.

The instructor should supply written evaluation reports to management outlining recommendations concerning Steps 1 and 2. The evaluation process should lead the instructor and management to answer the question, "What would we change, based upon what we know, to ensure that learning occurs in the easiest and most efficient way possible for students?" As can be seen, only human judgment can answer that question. In addition, that question can only be answered if criteria have been set in the form of course and specific behavioral objectives, and if evidence has been collected to support that the objectives have or have not been met.

SUMMARY

In any evaluation effort, three elements are critical to measuring success: criteria, evidence, and judgment. Criteria are the standards against which the instructor compares student learning after instruction. Criteria are most often found in the course behavioral objectives or the NFPA standards. They are the expected learner outcomes. Evidence is the information, data, or observation that allows the instructor to compare what was expected to what actually occurred. Judgment is the decision-making ability of the instructor to make comparisons, discernments, or conclusions about the instructional process and learner outcomes.

A judgment cannot be made without the presence of both of the other elements. Criteria alone only tell what was anticipated. Evidence alone only describes what occurred. But if learning is compared to the intended behavioral objectives, then an assessment can be made as to whether learning took place; and therefore, to the value of instruction.

Testing is an important part of evaluation and the instructor should be familiar with the different types of tests. Tests must be planned and the instructor should be familiar with the four steps of test planning. Tests may be classified into different categories.

One classification is based on the method of interpreting test results. Criterion-referenced and norm-referenced tests are the two tests under this classification. A second classification is based on the purpose of giving tests and the time at which they are given. Prescriptive, progress, and performance tests are under this classification. A third classification is based on the method of administration. Oral, written, and manipulative-performance tests are the tests found in this classification.

Statistics are used to analyze test results and help the instructor evaluate tests. The instructor should be familiar with the methods of analyzing test results. Testing students will enable the instructor to determine if the students have learned the required material.

Testing is just one part of evaluation. Evaluation includes both formative evaluation and summative evaluation. The instructor should use evaluation to determine how he or she can improve instruction and to uncover any problems in instruction.

SUPPLEMENTAL READINGS

Bloom, Benjamin S., George F. Madaus, and J. Thomas Hastings. *Evaluation to Improve Learning*. New York: McGraw-Hill, Inc., 1981.

Gronlund, Norman E. *Constructing Achievement Tests*. 3rd ed. Englewood Cliffs, New Jersey: Prentice-Hall, Inc., 1982.

Gronlund, Norman E. *Measurement and Evaluation in Teaching*. 5th ed. New York: Macmillan Publishing Company, Inc., 1976.

Mager, Robert F. *Measuring Instructional Results*. 2nd ed. Belmont, California: Pitman Learning, Inc., 1984.

Popham, W. James. *Criterion-Referenced Measurement*. Englewood, Cliffs, New Jersey: Prentice-Hall, Inc., 1978.

9

Computers in the
Fire Service

This chapter provides information that addresses performance objectives in NFPA 1401, *Fire Service Instructor Professional Qualifications* (1987), particularly those referenced in the following sections:

NFPA 1401

Fire Service Instructor

5-3.2 (n)

5-5

Chapter 9
Computers in the Fire Service

Computers have become widespread and are used for a variety of different applications. The fire service is no exception and many departments and fire service instructors are finding that computers are beneficial (Figure 9.1). Fire departments can use computers for word processing, record keeping, and training. The extent to which a department uses computers depends on many factors, including the size of the department and its financial resources. Some individuals are still leery of computers, but this attitude usually changes to acceptance when they see the benefits possible from using them. This chapter includes basic information on computers, different uses of computers, and how they can be applied to the fire service, particularly the fire service instructor.

Figure 9.1 Computers can be beneficial to the fire department and to the fire service instructor.

HISTORY

Microcomputers appeared in the public marketplace in the mid-to-late 1970s. The earlier, limited models have changed dramatically, and microcomputers are now capable of performing a variety of functions with relative ease. As microcomputers became popular they received a new label, personal computers, and have become powerful and efficient tools. The processing power available in personal computers today can be compared to the large, room-sized computers of the early 1960s.

As microcomputers became readily available, businesses and organizations realized they could benefit from using personal computers. Personal computers soon became commonplace in many organizations. The benefits of computers were easily realized and manufacturers began to market computers that were more user-friendly (Figure 9.2).

Figure 9.2 Microcomputers are now used widely and many organizations have found that they can benefit by using them.

INTRODUCTION TO MACHINES

Hardware

Hardware refers to the computer, electronic components, keyboard, CRT (monitor or screen), disk drives, and other physical items connected with the computer system (Figure 9.3). Each of these parts is significant to the system. The heart of a personal

Figure 9.3 Hardware refers to the physical items that make up the computer system.

computer is the central processing unit, or CPU. This is the part of the computer that actually processes information. Personal computers are rated by their processing speed, which is measured in megahertz (MHz). Many times personal computers are referred to by the amount of random access memory (RAM) they contain; 640 kilobytes (640K) being the maximum available in older PCs. Many newer personal computers offer 1 or more megabytes (1MB) of RAM as standard, with up to 32MB available for special applications. This is strictly electronic memory, not to be confused with disk storage capacity or mass storage.

MASS STORAGE
Floppy Drive

A minimum of one floppy drive is necessary to load the program onto the hard drive. Many word processors only require a single floppy disk for the program and a floppy disk for data storage. The introduction of supplemental word processing programs such as dictionaries, thesauruses, and graphics or large data bases will require a hard drive to hold all the programs or data. Floppy drives vary in capacities and size ranging from 5$\frac{1}{4}$-inch, 180K (kilobytes) capacity on older PCs to the 3$\frac{1}{2}$-inch, 1.44MB (megabytes) on newer PCs. The physical size of a diskette does not indicate how much information it can handle. The 3$\frac{1}{2}$-

inch disks, which are quickly gaining acceptance, have greater storage capacity than the standard 5¼-inch disks (Figure 9.4).

Figure 9.4 Floppy disks are used to store data.

Hard Drive

The hard or fixed drive can be thought of as several floppy disks stacked on top of each other. Hard drives allow a large quantity of information to be stored and readily accessible for the processor. Early hard drives had a capacity of 5MB, while current drives hold as much as 600MB. Capacities and access speeds change as technology progresses. The current recommended capacity in 1990 is 20 to 40MB for a basic unit.

Optical Disks

Optical disks are emerging as data storage devices. Although not suited for constantly changing data, they are well suited for data that remains constant or that changes slowly. Examples of this type of data are stored maps, chemical data bases, long-term health records, or long-term storage of employee training records. Optical disks resemble an audio compact disk and may hold up to 600MB of information on one disk. Some of the current training simulator software requires optical disks to store the simulation programs.

Software

Software is the program that performs a specific function or set of functions. Word processors, spreadsheets, desktop publishing, pre-incident planning, electronic mail, and data bases are but a few specific functions.

Software packages, usually called programs, are becoming more sophisticated, powerful, and easy to use (Figure 9.5). Almost all current software offers menus, help screens, or icons to assist the user. Icons are drawings or pictures that represent certain data, a program, or function.

Figure 9.5 Software programs are becoming more powerful, yet easier to use.

Software can be a sizable investment, depending on the tasks to be performed. Good programs are expensive to produce. Usually, a purchaser will buy hardware and give little consideration to software or will only budget for one software package such as the word processor. In reality, several different software packages may be used for instructional purposes. Software costs can mount to several hundreds or even thousands of dollars. A safe estimate for a budget would be to assume that software costs will be 75 to 100 percent of hardware costs. This may seem high at first, but different programs will quickly be accumulated as computer use increases. Instructors should realize that if software is copyrighted, making copies is a violation of copyright law.

If you already own a computer, the CPU and hard drive will dictate the type of software that may be used. Many desktop publishing applications require large amounts of memory, a high-speed processor, and a large hard disk for data storage. However, many word processors only require a single floppy disk for the program and a floppy disk for data storage. The introduction of supplemental word processing programs, such as dictionaries and grammar checkers, will require a hard drive to contain all of the extra programs.

The operating system is considered software. An operating system is an absolute requirement for any computer to function. The operating system takes instructions from other software, the user, or some input device and causes the hardware to react, in other words, an operating system is the interface between any set of instructions and the hardware. There are many versions of operating systems available with more being continually introduced. Some offer the user an easy-to-use interface of graphics or menus, while others only accept cryptic commands.

More complex operating systems are emerging to meet the needs of powerful programs, multiple users, and PC networks. As operating systems are becoming more powerful, however, developers are also attempting to make them easier to use, an important factor for novice users. Whether easy or difficult to use, the operating system is absolutely necessary.

SELECTING A COMPUTER SYSTEM

Anyone making a purchase for the first time must evaluate their needs (Figure 9.6). Even with little computer experience, training officers should be able to make an accurate assessment of their needs. First-time buyers should determine actual needs for the organization or individuals rather than relying solely on sales information. Fire departments considering a purchase might consult with other departments that already use computers.

Figure 9.6 Evaluation is an important step when planning a purchase.

The more information a buyer obtains before making a purchase, the easier that purchase will be. Otherwise, it is easy to be swayed by a sales pitch or a fast, colorful machine. Look at the tasks to be completed, and make a list of the department's present and future needs. Many communities have computer user groups that can provide assistance. The more the buyer knows about their needs, wants, and expectations the easier it will be to obtain the right kind of assistance.

Major User Costs

Major costs include just about everything bought during the initial purchase. A good computer system can be rather expensive. A basic minimum computer system should consist of a CPU, one floppy drive, a hard drive, a high resolution monitor, a good keyboard, and a printer (Figure 9.7).

At least one floppy disk drive is necessary to load programs onto the hard drive. Thus, many personal computers are being shipped with only one floppy drive. A high-resolution monitor is a must for long-term use, but it does not have to be a color monitor. Color is not necessary for word processing or record keeping, although it is nice for desktop publishing. Many training programs require a high-resolution color monitor for optimum performance. Regardless of the color, a sharp image is an absolute requirement for long-term users. User eye strain is a major consideration when making a purchase. Another consideration is the keyboard; it is important that the keyboard be comfortable to the user.

Printers are a requirement for any system. In many cases, dot-matrix printers will accomplish the job. However, near-letter-quality printers or laser printers will have much better type (Figure 9.8). Dot-matrix printers are relatively inexpensive and are good for a startup-type system. Depending on the usage, a printer that produces clear quality type may be a necessity.

Figure 9.7 Basic components of a computer system.

Figure 9.8 A laser printer gives a much better quality type but is also more expensive.

System Training

For the novice or even an experienced user buying a new product, training is essential (Figure 9.9). Usually, a computer store can supply training classes with an initial purchase. If this is not available, most community colleges offer short courses on using personal computers. In addition to basic PC use, courses on specific software packages are available. Fire departments should make use of available resources in their community. Even in smaller communities, there is probably someone familiar with computer systems.

Participating in a short course before purchasing a computer is a practical idea. It allows a prospective buyer to evaluate different hardware and software. This also enhances the user's knowledge when it is time to make a purchase.

Figure 9.9 Training is an essential part of obtaining a new computer system.

There are classes offered for operating systems just as there are for different software packages. A basic working knowledge of the operating system is imperative for the operator. While some operating systems offer a simple graphical interface, some are complex and cryptic. Having a working knowledge of your operating systems will be necessary for tasks such as formatting disks or installing software.

To introduce firefighters to computers, some trainers encourage the use of computer games as a recreational activity. As the users become more comfortable with the computer, training officers have noted that they will begin using other programs. Gradually, the user moves into application software and away from games. Using games as a teaching tool for the computer decreases the learning time required for computer training.

BENEFITS

Fire departments and instructors will benefit from using computers. Computers can be used for word processing, record keeping, and training simulations. Departments and instructors can use computers for desktop publishing, graphics, data analysis, and information exchange (Figure 9.10). Each of these applications can enhance an individual's or department's performance. Although these applications may not be labeled "fire service," each may be applicable to the fire service.

Word processors increase the efficiency in any situation that requires typed documents. In addition, record keeping on computers is widely promoted, a task for which computers are well suited. Computers can even be used as a form of communication. Teaching and presentation materials can be produced using different software applications, commonly called desktop publishing. Personal computers are even being used to create simulations for training purposes. Each of these areas can be used by the instructor and applied to the fire service.

Word Processing

Word processing is by far the most common application for a personal computer. Through the use of word processing and supplemental programs, people can create documents, make changes, check spelling and grammar, and even learn typing skills (Figure 9.11).

A word processor refers to the software package or programming that drives the personal computer, specifically for creating text documents. There are many of these packages available.

The instructor or anyone needing to create printed matter should become proficient with a word processor. Many are available with a variety of different features and in various price ranges. Some computer companies offer free word processing software with new computer purchases. Choose the word processing program that best suits your needs. Some programs are easy to learn, but may not offer a variety of editing options. Usually, those packages that offer the widest variety of options are more difficult to learn and may be more expensive. Some factors to consider when making a purchase are (Figure 9.12):

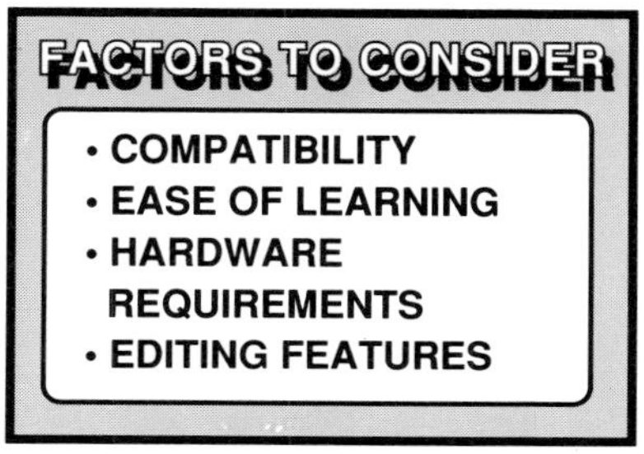

- Compatibility
- Ease of learning
- Hardware requirements
- Editing features

Compatibility, in this context, refers to the document or file produced by the word processor. Some factors that should be considered when looking at compatibility are:

- Will other word processors be able to read and edit the file?

COMPUTER USES

- WORD PROCESSING
- RECORD KEEPING
- TRAINING
- DESKTOP PUBLISHING
- GRAPHICS
- DATA ANALYSIS
- INFORMATION EXCHANGE

Figure 9.10 There are many benefits to using a computer.

WORD PROCESSING

- CREATE DOCUMENTS
- MAKE CHANGES
- CHECK SPELLING
- CHECK GRAMMAR
- LEARN TYPING SKILLS

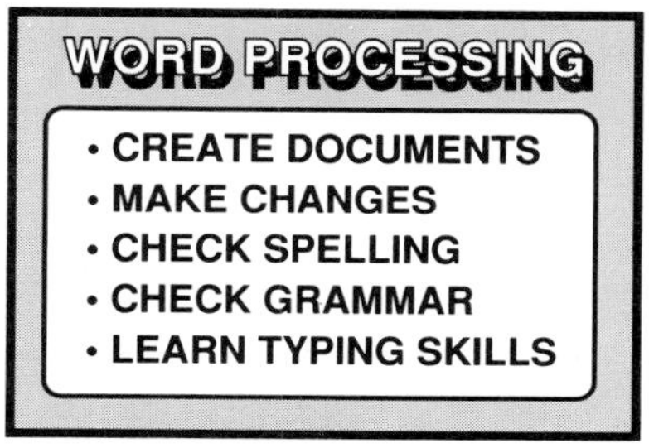

Figure 9.11 Benefits of word processing.

FACTORS TO CONSIDER

- COMPATIBILITY
- EASE OF LEARNING
- HARDWARE REQUIREMENTS
- EDITING FEATURES

Figure 9.12 Factors to consider when buying a word processing program.

- If you give your disk to another division of the department or another agency, will their computer and word processor be able to read the document?

- Also, will your word processor produce files that can be used later by other applications you wish to use, such as desktop publishing?

Purchase a program that is relatively easy to use. It takes time to become familiar with a new word processing program, but with practice and patience the program will become easy to use. Remember, also, that programs that offer more features may be more difficult to learn but might better suit your needs. Be prepared to spend time learning to use a new word processing program.

When shopping for a program look for the hardware requirements:

- If you already have a personal computer, will it fit in the computer's available memory?

- Does the program require a hard drive?

- Will the program support the printers you use or plan to use?

All of these questions should be answered before a purchase is made.

Pick a word processor that has good editing features. Obviously, a publishing firm has different requirements than an individual user. Look at the quantity and type of material you plan to create.

- Do you need a spelling checker, thesaurus, and grammar checker?

- Do you need to include graphics or spreadsheets into a document?

- Will you be using different types of printers?

All of these questions should be addressed. Usually, the more complex needs require a more expensive word processor.

Backups

When using a computer and keeping data on disks, certain precautions must be taken. Backup copies of information on disks should be made. A backup is a duplicate copy of your work. Many software companies offer programs that back up hard drives onto either floppy disks or tapes.

Backup copies of computer work should be kept away from your immediate work area. This assures that current work will not get confused with older work. Some organizations have fire-resistant vaults for keeping backups and original programs.

Sometimes duplicate copies are even kept in bank vaults, depending on the importance of the material. Generally, two backup copies are maintained. Only one copy is updated at a time. This way, if anything should happen to one set of backups, there is always another set for insurance. Although making backup copies may seem like a nuisance, it is a necessary precaution that should not be overlooked.

Documents

Word processing can be beneficial to the department and the instructor in helping prepare documents. Departments can use word processing for correspondence and when producing reports (Figure 9.13). The instructor can use word processing in a variety of different ways. Class schedules, outlines, correspondence, tests, and text can easily be generated. Word processors might be used to develop a teaching plan, or set up written notes for class handouts. Instructors can keep class notes on disk, making it easy to check what has been done during previous lessons.

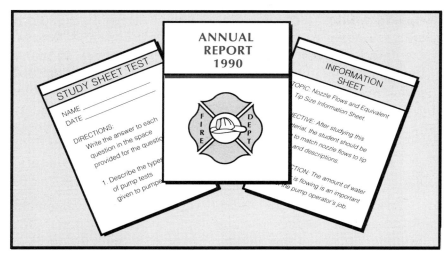

Figure 9.13 Word processing can be used for correspondence, class papers, and to produce reports.

Desktop Publishing

Desktop publishing makes it easy to develop training materials, overhead transparencies, and handouts (Figure 9.14 on next page). Materials are easily updated when developed on a computer. Additions and corrections may be printed out at any time rather than trying to update once a year. Using one of the slide creation programs, an instructor can create professional quality slides for presentations. This leads to a subject called desktop presentation; using a computer as part of the presentation rather than using it to create presentation material. This includes automated slide shows. Although some of these applications are not widely used, the instructor should not be limited in how he or she can use computers to enhance teaching.

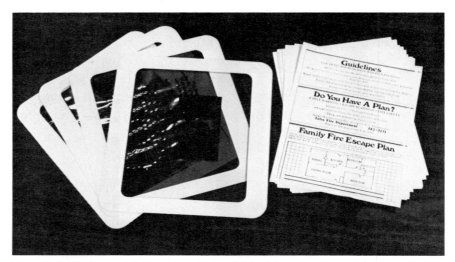

Figure 9.14 Desktop publishing can aid in developing instructional materials.

Record Keeping

Computers can be used for record keeping (Figure 9.15). The types of programs used for record keeping generally fall into the group called data bases, of which many variations exist. Record keeping is a generic term that describes a data base and its uses. These data bases can be used for statistical analysis, in many cases either from in-house information or from state and national organizations. There are many different data bases available. These, like word processors, can range from simple to complex. Some companies offer data base programs for specific purposes that are structured for a specific task. Some prime uses for data bases are to keep track of training records, equipment maintenance, incident reports, and material safety data sheets.

Before purchasing data base software, the department should again assess its needs. Does the department need a complex program just to keep up with training records? Shop for existing programs that perform the needed functions. If the department

Figure 9.15 Computers can help a department with its record keeping.

elects to purchase data base software and develop its own programs, make sure the department has a workable filing system. Create a filing system that works with paper or a card file before attempting to automate it. If the system does not work on paper, the computer will only make it worse!

Training Records

Personnel training records are excellent items to computerize. Class, departmental, and individual training records can be computerized. Many companies offer software for training records in canned programs. These programs are usually modular, meaning that they may be integrated into other software packages sold by the same company or operated as a stand-alone program.

MANUAL TO ELECTRONIC

If the department moves to an electronic record keeping system, be prepared for problems. **Do not totally rely on the computer for record keeping until you are thoroughly familiar with the operation.** It is wise to maintain training records manually (on paper) for at least six months to a year after implementing the computer. This will ensure that the department has control over record keeping if data on the computer is lost, or improperly entered.

Security

Security is an important issue when training or financial records are kept on a computer. Software for training or financial records should contain some measure of security. Training and finances are a private matter and should not be readily accessible to unauthorized users.

Some programs offer built-in security that requires passwords before the user can gain access to any information. If this is not available, the data should be kept on a floppy disk and locked in a cabinet or separate room. Another way to restrict access to sensitive data is to lock the computer in a secure room to which only a few people have keys.

Graphics

Personal computers may also be used for creating graphics. The term graphics covers many different applications, ranging from simple drawings to complete publications. Software required to produce more detailed publications, such as this manual, can be complex, difficult to learn, and require specialized equipment. However, there are many programs available that work well on less expensive computers that allow users to create such things as monthly newsletters, lesson plans, handouts, overhead transparencies, and slide programs. In short, evaluate your needs before acquiring such programs and/or expensive equipment. Even the less expensive programs require practice. The users should allocate enough time to become proficient with any of these programs to avoid frustration.

Most of these programs are expensive, particularly if they are not used often. In general, only organizations that produce large amounts of graphic materials can usually justify the costs involved for complex electronic publishing programs and equipment. In any event, a laser printer is a must for producing good graphics, which adds more costs to the overall system. The instructor may want to find a local graphics company and have certain materials prepared for the class rather than prepare the material themselves. Some communities may have businesses that prepare reports and do typing that may also be able to do some graphics. Generally, the computer will enhance the abilities of anyone wishing to develop graphical material. The instructor should try several different software applications prior to purchasing one if possible.

For graphic presentations to a group, an instructor may elect to use an overhead projector coupled to a projector screen for the computer. This device allows the image seen on the computer screen (CRT) to be projected just as any transparency. This device will certainly enhance a presentation, although using it requires patience and practice. The instructor should remember rules regarding enlarging text. The character on a computer screen does not lend itself to enlargement due to the font type, weight, and size. Students may be viewing small illegible characters on an overhead that are quite sharp on the instructor's computer screen.

Computer-Aided Instruction (CAI)

The use of computers in education and training is still a new and rapidly evolving field. CAI requires a relatively large initial investment, but a good CAI program can offer several advantages over more traditional instruction techniques for fire department training programs.

A well-designed CAI program offers flexibility in scheduling, learner pacing, and student-computer interaction that enhances learning efficiency and effectiveness (Figure 9.16). The extent of interaction between the learner and the computer varies among programs. In some programs, the computer simply presents the workbook-like lessons in a specified order. In others, the computer provides additional instruction on selected topics, based on the performance of the student. In addition, some simulation programs present situations that change and evolve depending on the operator's input.

Because CAI is a relatively new field, the training material available is constantly changing. One alternative is to use pro-

CAI OFFERS

- FLEXIBILITY IN SCHEDULING
- LEARNER PACING
- STUDENT-COMPUTER INTERACTION

Figure 9.16 A CAI program can provide benefits that will enhance learning.

grams that have been developed by others for training. It is possible to find training programs that have been specifically designed for training use in the fire service.

Another alternative is to develop your own training programs. Trying to program a computer to present the lessons is often a frightening idea. Authoring programs are available that allow instructors to put together CAI programs with a minimum of programming skills.

Examination programs are good for administering standardized tests and scoring students equally. Many programs allow the instructor to add test questions to an already existing bank of questions. Instructors may also be able to structure tests that weigh heavily on one particular subject.

COMPUTER SIMULATORS

Computer simulators are a relatively new method of training and have yet to reach even a fraction of their potential, especially in the fire service. The most sophisticated computer simulators can be found in the military and commercial airline sectors.

Simulators for the fire service generally consist of model buildings for tactical training or pump panel mockups for operator training. These items usually do not have anything to do with a computer, but they are the predecessors for full-scale, computer-based simulators. Computer programs for fire training are still in the developmental stages. The products on the market generally are relatively expensive, but for the most part are not highly sophisticated. Even so, these are still useful items when teaching students at a self-paced schedule. Most self-paced programs allow the instructor more flexibility when assisting students. Generally, those students that need to spend longer time on certain material may do so without impeding the class.

Data Analysis

Personal computers can be used for analysis of data in many different situations. Personal computers allow some departments to use a computer to complete incident reports. A popular way to analyze large amounts of data is the use of database programs in conjunction with computer spreadsheets. The databases are capable of manipulating large amounts of data and exporting specific information to different spreadsheet programs. This feature allows a user to manipulate information or even graph it. There are also specific computer programs for such things as statistical analysis (Figure 9.17 on next page).

Personal computers have evolved to the point where they can share information with mainframe computer systems. The National Fire Incident Reporting System (NFIRS) takes incident reports from fire departments and compiles national statistics.

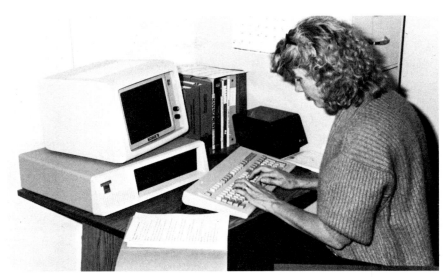

Figure 9.17 Computers can aid departments in data analysis.

These reports are usually handwritten by firefighters after an incident and must be completed in a very specific manner. Advances in computer technology now allow many fire departments to buy software that allows a firefighter to complete NFIRS or other reports on a personal computer. This software usually assists in placing proper codes and descriptions in the report. Once the fire reports are filled out, they may be kept in a data base for statistical review. At the end of the reporting period when these reports are turned over to the national level, either the disks are sent in or they are transmitted electronically across telephone lines. This is also true for local and state level reporting. Where states have an incident reporting system, they usually accept reports in electronic format.

NOTE: Check with the agency involved before sending any information in electronic form since some formats may not be compatible.

A byproduct of compiling incident reports into electronic form is the ability to retain and use these reports for statistical analysis on a local level. Some of the current reporting software allows individual fire departments to develop reports and analyze trends from their own incidents. All of this simplifies work and reduces errors in reports.

Spreadsheet Applications

A popular application of microcomputers is the use of spreadsheets. Spreadsheets are computer programs that are capable of manipulating data or numbers. A typical spreadsheet program consists of columns and rows. The first spreadsheets were used mainly in business for such tasks as accounting ledgers, which they closely resemble on the screen, but they soon found their way into other areas of computer use. Spreadsheets permit many numerical tasks to be done much easier than could be accomplished by hand; for example, determining budgets.

Information Exchange

By using a device called a modem, a computer can communicate with other computers over telephone lines. Electronic mail is becoming a popular and efficient way of communicating. In addition to electronic mail, electronic information systems are available. These systems allow a person to search data and records for a fee.

Information may be exchanged through computer bulletin boards. Computer bulletin boards contain information about a specific subject and function much like their physical counterparts. An individual computer user posts a message about a particular subject or question. Other computer users access the board at different times and read the available messages. If anyone has information to share, they may leave it for a specific individual or make it public for everyone. Many informal fire-related bulletin boards exist. The International Association of Fire Chiefs is sponsoring a commercial bulletin board and electronic mail service called ICHIEFS. For a fee, any individual or fire department may purchase software that will communicate through this system. In addition to electronic mail, items such as pending legislation, articles on major incidents, arson, and training information can be found on ICHIEFS.

INTERDEPARTMENTAL COMMUNICATION

Some fire departments, both large and small, are using computer networks for nonemergency communication between stations. This may be part of a computer-aided dispatching system, such as the one used by the Phoenix (Arizona) Fire Department, or an isolated network like that used by the Santee (California) Fire Department. These systems allow a computer in each station to pass information along to other computers, thus allowing each station to access training records, incident reports, and news bulletins as they are updated.

SUMMARY

Computers can be beneficial to both the department and the instructor. Computers consist of hardware and software. Hardware refers to the computer, electronic components, keyboard, CRT, disk drives, and other physical items connected with the computer system. Software consists of the programs that enable the computer to perform specific functions.

Before purchasing a computer, departments should assess their needs. A good computer is an expensive investment and departments should evaluate what they want out of a computer system. Training is an essential part of obtaining a computer system, and people must be able to use the computer for it to be effective.

Computers can be used for a variety of different applications in the fire service. Word processing, desktop publishing, record keeping, and graphics are some specific uses that can benefit the instructor.

The fire service instructor can help prepare lessons and instructional materials on the computer. Computer-aided instruction and computer simulators are two relatively new advances in teaching that have potential for wider use in the fire service. Departments can also use computers for data analysis, information exchange, and interdepartmental communication.

SUPPLEMENTAL READINGS

Kearsley, Greg. *Computer-Based Training.* Reading, MA: Addison-Wesley Publishing Company, Inc., 1984.

Kearsley, Greg. *Training for Tomorrow.* Reading, MA: Addison-Wesley Publishing Company, Inc., 1985.

Lockard, James, Peter D. Abrams, and Wesley A. Many. *Microcomputers for Educators.* Boston: Little, Brown and Company, 1987.

Siegel, Martin A., and Dennis M. Davis. *Understanding Computer-Based Education.* New York: Random House, 1986.

White, Charles S., and Guy Hubbard. *Computers and Education.* New York: Macmillan Publishing Company, 1988.

Heinich, Robert, Michael Molenda, and James Russell. *Instructional Media and the New Technologies of Instruction.* New York: Macmillan Publishing Company, 1989.

Appendices

Appendix A

SAMPLE LESSON PLANS

Appendix A presents a preview of a professionally developed curriculum design. The sample user's guide, lesson plan, and application activities are typical of the competency-based curriculum presently being developed by Fire Protection Publications for a number of our IFSTA validated fire service manuals. The copyrighted format, soon to be on the market, contains each of the four elements basic to the development of any effective curriculum design: preparation, presentation, application, and evaluation.

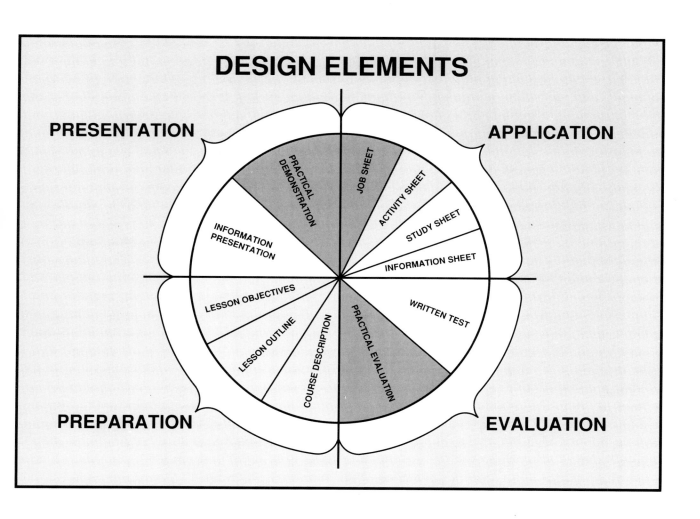

DESIGN ELEMENTS

PRESENTATION

APPLICATION

PRACTICAL DEMONSTRATION

JOB SHEET

ACTIVITY SHEET

STUDY SHEET

INFORMATION PRESENTATION

INFORMATION SHEET

LESSON OBJECTIVES

WRITTEN TEST

LESSON OUTLINE

COURSE DESCRIPTION

PRACTICAL EVALUATION

PREPARATION

EVALUATION

IMPLEMENTATION GUIDE

INTRODUCTION

Successful users of the curriculum materials for **Company Officer** must first have an understanding of the course structure and the materials that will be used. This implementation guide is designed to provide information on the following subjects:

- Course Structure
- Content Overview
- **Instructor's Guide** Materials
  - —Implementation Guide
  - —Lessons
  - —Lesson Tests/Answers
  - —Content Review Test Answer Template
  - —Transparencies
- **Student Guide** Materials
  - —Text
  - —Study Objectives
  - —Information Sheets
  - —Study Sheets
  - —Activity Sheets/Answers
  - —Practical Activity Sheets/Evaluations
  - —Job Sheets/Performance Evaluations
  - —Content Review Tests/Answer Sheets
  - —Competency Profile
- Teaching Tips

COURSE STRUCTURE

This course is structured to assist both competency-based group instruction and self-study of the materials presented in the IFSTA validated **Company Officer** manual, Second Edition.

Like a wheel, the course revolves around the course objective: *to teach the student the information and skills needed to qualify for the fire service position of company officer*. Radiating out from this hub, like the spokes on a wheel, are the lesson plans—each containing listed competencies (objectives) that precede and shape each lesson.

To enable the instructor to teach listed competencies, detailed lesson plans keyed to corresponding chapters and main points in **Company Officer** are contained in the **Instructor's Guide**. A written test is included with each lesson so that the instructor can assess whether or not the student has mastered the cognitive lesson objectives.

To enable the student to master the listed competencies, application and self-study activities are presented in the companion **Student Guide**. Activities in this guide allow the student to practice, apply, and demonstrate competency in each of the psychomotor objectives. Practical and performance tests are included to evaluate student mastery of these objectives.

In addition to addressing the psychomotor objectives, student materials also contain supplementary information and activities that reinforce classroom instruction, as well as study guides and activities designed for self-instruction.

CONTENT OVERVIEW

Individual lessons in the **Company Officer Instructor's Guide** contain the following content. (**Company Officer** lesson numbers and the chapters they correspond to are referenced in parentheses.)

SECTION I—FITTING INTO THE ORGANIZATIONAL SCHEME

- **Introduction—Company Officer: The Vital Connecting Link** *(Lesson 1, Introduction)*

Understanding the company officer's expanding role in the fire service; evaluating personal learning objectives and career goals for the course; learning rules for success as a company officer; discussing problems encountered in the transition from firefighter to officer.

- **Learning the Principles of Organization** *(Lesson 2, Chapter 1)*

Distinguishing between line and staff; understanding the principles of unity of command, span of control, and division of labor; understanding the concept of discipline.

- **Understanding Fire Service Organizational Structure** *(Lesson 3, Chapter 2)*

Learning the basic forms of company and departmental organizational structures and the company officer's position in a scalar flowchart; distinguishing between centralized and decentralized authority; understanding the relationships between line and staff personnel.

SECTION II—HANDLING INTERPERSONAL RELATIONSHIPS

- **Communicating Effectively** *(Lesson 4, Chapter 3)*

Developing effective verbal and written communication skills; learning the importance of effective face-to-face communications on the fireground.

- **Learning Group Dynamics and Effective Group Interaction Skills** *(Lesson 5, Chapter 4)*

Distinguishing between formal and informal groups; learning the five elements of all groups; understanding Maslow's five levels of need and how they relate to the company; recognizing factors that act as determinants of effective group interaction.

- **Influencing the Group Through Leadership** (*Lesson 6, Chapter 5*)

  Learning the types of leadership; evaluating the demands of different leadership styles; analyzing the types of power used by leaders; evaluating individual leadership potential and style.

- **Filling the Role of Manager** (*Lesson 7, Chapter 6*)

  Learning the management cycle; using the MBO system; learning the importance of evaluating employee performance; establishing goals.

SECTION III—MANAGING INDIVIDUAL PERFORMANCE

- **Motivating Employees** (*Lesson 8, Chapter 7*)

  Recognizing positive results; giving constructive feedback; tapping the diverse resources of all team members; learning influences and needs.

- **Career Counseling** (*Lesson 9, Chapter 8*)

  Learning the basis of employee evaluation; evaluating employee performance; learning acceptance standards for various fire service positions; determining possible courses of action for career goals achievement.

- **Taking Corrective Action** (*Lesson 10, Chapter 8*)

  Dealing with emotional behavior; dealing with conflict; correcting unproductive behavior; applying the progressive system to properly discipline an employee.

- **Solving Problems** (*Lesson 11, Chapter 9*)

  Using the eight-step problem-solving process; participating in a team problem-solving session; encouraging commitment to cooperative solutions; handling complaints; introducing change.

- **Mid-Course Content Review Test** (*Intro - Chapter 9*)

SECTION IV—MANAGING PRE-INCIDENT AND FIREGROUND PROCEDURES

- **Making Pre-Incident Surveys** (*Lesson 12, Chapter 10*)

  Distinguishing between pre-incident surveys and pre-incident planning; determining the role of the company officer as a member of the pre-incident survey team; performing a pre-incident survey.

- **Managing Fireground Procedures** (*Lesson 13, Chapter 11*)

  Distinguishing among strategic plans, operational strategies, and tactics; determining attack modes and operational strategies; learning how the tactical development process affects operational strategies; developing a strategic plan.

- **Performing Size-Up** (*Lesson 14, Chapter 12*)

  Learning the major considerations in size-up; performing size-up on a structure; describing the relationship between the incident and available resources.

- **Learning Incident Command and Communications Skills** (*Lesson 15, Chapter 13*)

  Using fireground command and communications procedures; assigning command; transmitting effective radio communications.

SECTION V—ADMINISTERING SAFETY, HEALTH, AND LEGAL GUIDELINES

- **Learning Firefighter Safety and Health Practices** (*Lesson 16, Chapter 14*)

  Learning NFPA 1500 guidelines under the company officer's direct supervision; identifying the most frequent causes of injuries and fatalities to firefighters; identifying potentially hazardous conditions; recognizing various levels of stress; identifying company members who may be substance abusers; learning the purpose and advantages of a wellness program; counseling firefighters in methods of stress reduction; investigating and analyzing company time-loss injuries.

- **Knowing the Company Officer's Liability** (*Lesson 17, Chapter 15*)

  Learning the principles of administrative law; distinguishing between legal duty and liability; defining tort liability and understanding its components; knowing the definitions of negligence, standard of care, and causation; learning the elements of due process; and explaining the company officer's responsibility for subordinates' actions and role in the affirmative action process.

- **Final Content Review Test** (*Lessons 1-17, Intro – Chapter 15*)

INSTRUCTOR'S GUIDE MATERIALS

The **Company Officer Instructor's Guide** provides the instructor with a variety of materials designed for effective, competency-based instruction and ease of delivery. Instructor's materials fall into four groups: Implementation Guide, Lessons, Lesson Tests/Answers, and Transparencies.

IMPLEMENTATION GUIDE

The Implementation Guide is what you are reading right now. "Implement" means tool, and this implementation guide should help you become familiar with the various tools or components of both the instructor and student guides. Look at each component as you read about it to familiarize yourself with its format. Note that each page lists the course component as well as the page number. This system allows you to remove and reinsert pages easily.

LESSONS

Each **Instructor's Guide** contains lessons referenced to chapters and main points in the IFSTA validated **Company Officer**. The lessons include specific directions for what to do and say, yet they are flexible enough to allow you to personalize and individualize instruction. Each lesson is laid out in the following order:

- *Title Page*—identifies the lesson.
- *Objective Page*—appears on the back of the title/divider page and contains—
  1. *Course Prerequisite*—identifies courses or training that the student must master before taking this course

  2. *Lesson Prerequisite*—indicates lessons that must be taught (learned) before this one; is used to help the instructor present the course in the correct sequence, and to allow the student to build upon learned material

  3. *Course Objective*—(first lesson only) indicates the instructional goal of the course

  4. *Lesson Objective*—indicates the instructional goal of the lesson

  5. *Enabling Objectives*—indicate the specific performances necessary to achieve the lesson objective

  6. *Lesson Outline*—lists subject blocks covered in the lesson and time frames for each subject and for total class session; estimated times are given for teaching the lesson with or without teaching options

- *Planning Pages*—provide teaching and preparation suggestions, as well as a list of transparencies, flipcharts, handouts, activity and job sheets included in the lesson. These pages also include references used in the development of the lesson. You may use these to supplement your knowledge of the subject or to help students with particular interests or occupational objectives in the subject area.

- *Step-by-Step Lesson Plan*—provides information and detailed directions on how to present material needed to accomplish the lesson objectives.

- *Handouts*—supplementary materials such as self-evaluations, session guides, charts, and motivational materials. They are designed to be copied and distributed to individual students or to the class as a whole.

- *Transparency Summary*—visual index of all transparencies that must be selected before presenting the lesson.

- *Flipchart Summary*—visual index of all flipcharts that must be prepared before delivering the lesson.

- *Lesson Test*—short-answer, criterion-referenced test over the lesson's cognitive objectives.

Typical Lesson Activities

A typical lesson within the **Company Officer Instructor's Guide** includes the following:

- *Administration*—routine classroom duties such as taking roll, collecting assignments, distributing handouts, etc.

- *Review*—a summary review of the main points covered or the skills acquired in the previous lesson; often a critique of the lesson test administered in the previous lesson.

- *Motivation/Introduction*—combined or separate, these two activities are designed to catch the student's interest and present the topic of the lesson; may include a combination of "war stories," quotation discussions, and visuals that focus the student on the importance and relevance of the material or skill being taught.

- *Content Presentation*—provides information necessary to attain the lesson objectives; material is reinforced with visuals, written exercises, and practical applications.

- *Teaching Options*—supplemental teaching ideas that you may choose when reinforcement and repetition are needed. Teaching options are presented in colored type and are not tested objectives.

Lesson Format

Lesson pages are divided into three columns:

Cue	*Content*	*Notes*
Provides visual symbols and action verbs to direct the instructor to major activities.	Includes quoted statements, information, and questions; displays suggested flipcharts and transparencies.	Provides space for instructor's notes and personalization of lesson plan.

Cue Column: Visual Cues
Visual symbols (many containing the reference number of the needed visual) are used in the cue column for ease of reference. The instructor need only glance at the lesson to know the activity or course of action to take.

Here is an index of the visual symbols used in the lesson cue column:

Refer students to **Student Guide** or text. Display transparency.

Use Teaching Option or continue with standard lesson.

Distribute handout, or return homework or test.

Use prepared flipchart.

Show film or slide/ tape presentation.

Show videotape.

Ask question.

Refer student to lesson test.

Make an assignment.

Cue Column: Verbal Cues

The cue column also provides directions in the form of action verbs. The following action verbs are used most often in the cue columns of each lesson.

Ask—Ask question(s).

Explain—Talk *to* students; tell what, how, and when; give examples, answer questions, clarify.

Discuss—Talk *with* students; exchange views to find out "why"; ask questions to involve students, provide examples, draw conclusions.

Display—Show transparency or object.

Post—Hang flipchart page where students can see and refer to it.

Distribute—Provide students with handout, review questions, homework, or other classroom material.

Administer—Give Lesson Test, Practical Evaluation, Performance Evaluation, or Content Review Test.

List—Record responses on flipchart or chalkboard.

Review—Go over material covered earlier.

Refer—Turn to (or have students turn to) specified flipchart, page in text, or component in **Student Guide**.

Summarize—Briefly review main points of lesson or subject block.

Content Column

Information in this column corresponds with the lesson's objectives and activities. It is divided into sections, or blocks, that cover material within one subject area or objective. A horizontal box containing the section subject and suggested time frame separates each subject block.

The content column displays flipcharts and transparencies for ease of reference. Flipcharts that end with a dotted bottom line are ones on which you will write student responses.

```
REASONS FOR REVIEW
```

Flipcharts with a solid bottom line are complete as is and do not require space for student responses.

```
REASONS FOR REVIEW
Memory aid
Preparation for evaluation
Continuity
```

Notes Column

This column allows you to personalize each lesson with your own notes, materials, examples, or instructions.

You are urged to personalize and localize the curriculum, adapting it to meet city, county, state, or province requirements for certification.

Here is a sample page:

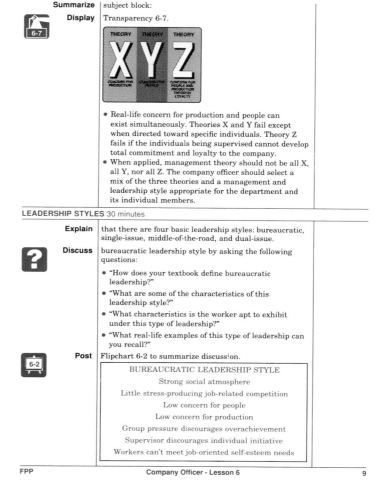

Summarize subject block:

Display Transparency 6-7.

- Real-life concern for production and people can exist simultaneously. Theories X and Y fail except when directed toward specific individuals. Theory Z fails if the individuals being supervised cannot develop total commitment and loyalty to the company.
- When applied, management theory should not be all X, all Y, nor all Z. The company officer should select a mix of the three theories and a management and leadership style appropriate for the department and its individual members.

LEADERSHIP STYLES 30 minutes

Explain that there are four basic leadership styles: bureaucratic, single-issue, middle-of-the-road, and dual-issue.

Discuss bureaucratic leadership style by asking the following questions:

- "How does your textbook define bureaucratic leadership?"
- "What are some of the characteristics of this leadership style?"
- "What characteristics is the worker apt to exhibit under this type of leadership?"
- "What real-life examples of this type of leadership can you recall?"

Post Flipchart 6-2 to summarize discussion.

BUREAUCRATIC LEADERSHIP STYLE
Strong social atmosphere
Little stress-producing job-related competition
Low concern for people
Low concern for production
Group pressure discourages overachievement
Supervisor discourages individual initiative
Workers can't meet job-oriented self-esteem needs

FPP — Company Officer - Lesson 6 — 9

LESSON TESTS/ANSWERS
Multiple-choice, matching, true/false, and short-answer lesson tests (LT), located at the end of each lesson plan, should be used to evaluate the accomplishment of the cognitive lesson objectives. Answers are printed on a separate answer sheet. Lesson tests should be copied and administered as soon after teaching the lesson as is practical.

CONTENT REVIEW TEST ANSWER TEMPLATE
This template has been designed for ease of scoring the multiple-choice content review tests. The A, B, C, and D designations in the corners of the template indicate the corresponding test/answer sheet. For example, if the students have taken review test A, place the template on the A test answer sheet with the A in the upper right-hand corner. The correct answers will show through the template holes. If there is no answer in a hole, the test item is marked as incorrect.

Answers to tests B, C, and D are scored in similar fashion, their template letters aligned in the upper right-hand corner over the answer sheet.

TRANSPARENCIES
Location and Access
Transparencies (T 1-1, T 1-2, and so on) for the entire course are collated and packaged separately for use with the **Instructor's Guide**. Overlays are indicated by small letters (T 1-1a, T 1-1b, etc.). A sturdy three-snap frame is included in the transparency package, and should be used to support the three-hole-punched transparencies while they are on the overhead projector.

You may choose to select only those transparencies you need for a specific lesson, or you may find it convenient to refer to the lesson transparency summaries and insert the transparencies with their corresponding lessons.

How to Use
Follow the teachings of Confucius, who wrote:

I hear and I forget.

I see and I remember.

I do and I understand.

Use the transparencies for motivation, and to reinforce, summarize, and emphasize. Showing a transparency is often an excellent way to introduce a topic or initiate a discussion. You can also mark directly on a transparency to fill in blanks or charts, thus involving students more directly, or you can present the subject a little at a time by covering material you do not want projected. In short, let the students "see and remember," but be sure to use transparencies as supplemental teaching aids and not as the primary teaching method. The student will "hear and forget" if the instructor displays transparency after transparency and merely reads the copy on them.

STUDENT GUIDE MATERIALS

Students should obtain a copy of the IFSTA validated **Company Officer**, Second Edition, copyright 1990, and a copy of **Company Officer Student Guide**, First Edition, copyright 1990.

Copies of each are available from:
Fire Protection Publications
Oklahoma State University
Stillwater, Oklahoma 74078-0118
1-800-654-4055

The following components are listed in the order in which they appear after each lesson/chapter block in the **Student Guide**.

Each component is keyed to its corresponding lesson or chapter to make it appropriate for group as well as self-instruction.

The first activity for Lesson 6, for instance, would be numbered AS 6-1, the second, AS 6-2, and so on. Answer sheets are provided when appropriate.

STUDY OBJECTIVES

Study objectives (SO) tell the student what he or she is expected to know as a result of studying the chapter and completing related activities in the **Student Guide**. Study objectives for each subject block are listed on the backs of the divider pages. These objectives correspond specifically to the lesson's enabling objectives and generally to the text's chapter objectives.

INFORMATION SHEETS

The information sheet (IS) contains supplementary material related to but not found in the course text. Information sheet material grows out of one of the lesson's cognitive objectives, and as such is tested in the Lesson Test.

STUDY SHEETS

This component of the **Student Guide** is designed basically for self-instruction, but may also be used as a study guide by students of group instruction. Study sheets (SS) contain vocabulary, chapter and subject questions, case studies, supplementary assignments, and other activities that help students apply concepts and learn material they may be tested on.

ACTIVITY SHEETS/ANSWERS

Activity sheets (AS) provide students with pencil and paper activities that allow them to practice cognitive skills or to complete affective activities. Activity sheet exercises may involve the student in problem-solving, analysis, self-assessment, appropriate practice, or other applications of cognitive or affective material. Answers are provided where applicable.

Because activity sheets are designed for practice or self-evaluation, they are not listed in the lesson's objectives nor evaluated or tested.

PRACTICAL ACTIVITY SHEETS/EVALUATIONS

Practical activity sheets (PAS) present students with activities that assist them in applying cognitive and manipulative skills that result in a product. Practical activity sheets may include such activities as writing a report, preparing and giving an oral presentation, filling out forms, creating a budget, writing letters and memos, and responding in writing to case studies or scenarios. These activity sheets are listed as enabling objectives, and as such are evaluated competencies. They are not evaluated on the written test, however, but are evaluated independently with a Practical Evaluation.

The practical evaluation that accompanies each PAS emphasizes and assesses the *product* of the student's endeavors rather than the procedure—as in a job sheet. The

PAS product rating is converted to a competency rating and recorded on the student's competency profile. Students must achieve a minimum competency rating of 3 (moderately skilled) to show mastery of each PAS objective.

JOB SHEETS/PERFORMANCE EVALUATIONS

Job sheets (JS) are valuable class resources. They provide students with a list of equipment, tools, and materials and a step-by-step procedure for learning a manipulative or psychomotor skill.

Job sheet procedures are used by the instructor for practical demonstration of psychomotor skills and are designed for student practice, not for self-instruction. You should demonstrate the procedures outlined in the job sheets, and then allow students to practice these procedures *under your supervision*. You must have the student's safety in mind at all times.

Job sheets are listed as enabling objectives and are very important fire service competencies.

After the student has practiced enough to learn the psychomotor skill, job sheet competencies are tested with a performance evaluation. Administered by the instructor or an authority having jurisdiction, these performance tests assess *performance*—student achievement of the lesson's psychomotor objectives—the practical skills covered in the job sheet. Performance test results should be recorded on the competency profile.

CONTENT REVIEW TESTS

These two comprehensive evaluation instruments—mid-course and final review tests—are collated at the end of the **Student Guide**. Answer sheets are included with each to facilitate scoring.

These tests are designed to be used by either the instructor or the student. For instance, you may use the tests to evaluate retention of information presented in the course or text, as a pretest, or to reinforce material presented in a lesson. Students may use the tests as study or review aids.

COMPETENCY PROFILE

The competency profile is an instrument for documenting student achievement levels for each of the course objectives.

At the beginning of the course, students complete the cover of the profile and then give it to you for documenting competency levels and achievement throughout the course.

TEACHING TIPS

The instructor's package, which contains the **Instructor's Guide**, transparencies, and the **Student Guide**, is designed to provide you with everything you need to teach a successful course. But it is the teacher, not the materials, that ultimately makes a course successful. As a user of this curriculum, you can improve your success rate by following these suggestions:

- Adequately prepare for each lesson.

  1. Read the lesson plans and prepare your own notes, examples, and supplementary materials.

  2. Study the lesson and study objectives. Know the goals for the lesson and how you (and the students) are going to achieve them.

  3. Read the activity, job, study, and information sheets so you can anticipate student problems and set up needed equipment. Modify job sheets as necessary to fit available equipment and local needs. Schedule performance evaluation times and dates.

  4. Prepare flipcharts needed to facilitate class discussion. Be creative! Use your imagination and colored markers to "jazz up" the flipcharts and wake up the class.

  5. Use the transparencies and visuals as teaching aids, not as teaching method; and arrange in advance for projectors, recorders, and VCR equipment. Know how to operate all audiovisual equipment and be sure it is in working order before you present the lesson.

  6. Study the session guide and personalize it to fit your specific teaching needs. Write in page numbers, supplementary reading assignments, due dates, and classroom rules and policies.

  7. Review teaching options and decide which you will include in your lesson.

  8. Administer lesson tests and performance evaluations as soon after teaching the lesson as possible. Score promptly and document on competency profile.

- Follow the cues for delivering each lesson, and pace yourself to stay close to the suggested delivery times.

- Keep the lecture to a minimum; remember the words of Confucius, and actively involve the students in the learning process.

- Ask open-ended questions that help draw out students' ideas and experiences.

- Be enthusiastic!

LESSON 6

CHAPTER 5
COMPANY OFFICER

For *ifsta* Company Officer

LESSON 6 — INFLUENCING THE GROUP THROUGH LEADERSHIP

PREREQUISITE

Lesson 5—Learning Group Dynamics and Effective Group Interaction Skills

OBJECTIVES

Lesson: After completing this lesson, the student will be able to recognize the need for different leadership approaches and will be able to apply these leadership skills to the fire service position of company officer.

Enabling: After completing this lesson, students should be able to—

1. Define *leadership.*

2. Recognize the three theories of leadership.

3. Distinguish among leadership styles.

4. Describe the types of power used by leaders.

5. List the five characteristics of effective leadership.

6. Organize and lead a group session. (Job Sheet 6-1)

LESSON OUTLINE

Topic / Activity	Est. Time
Administration/Review	10 min.
Introduction/Motivation	10 min.
Leadership Theories	15 min.
Leadership Styles	30 min.
Teaching Option 6-1: Activity Sheet 6-1—Analyze Leadership Styles	20 min.
Power Structures	15 min.
Teaching Option 6-2: Power Play	15 min.
Characteristics of Effective Leadership	20 min.
Summary: Rise to New Heights of Leadership	15 min.
Lesson Test	30 min.
	TOTAL 2 hrs., 25 min.

TOTAL WITH TEACHING OPTIONS 3 hrs.

LESSON 6—PLANNING PAGES

- Score Lesson 5 Test and record on competency profile.
- Read **Company Officer**, Chapter 5.
- Read and familiarize yourself with this lesson. Prepare your own notes and examples.
- Refer to Transparency Summary at end of lesson. Review transparencies and stack in correct teaching order.
- Write flipcharts by referring to Flipchart Summary at end of lesson. Be sure to leave blank pages after those that require student input.
- Duplicate appropriate number of Lesson 6 Test.
- If you plan on using the Power Play Teaching Option 6-2, prepare fifteen 3x5 flashcards by writing each of the following on a card. Shuffle the cards well when you are finished.

Front	Back
"Jim, you did a super job with your housekeeping duties. I've never seen the kitchen look better."	Reward power
"Good morning Tom. It's great to see you back and looking fit."	Reward power
"And I think that we should all give Pete a round of applause for the planning and hard work he put into yesterday's exhibition."	Reward power
"For Pete's sake, Josh, you seem to be all thumbs. Anyone with half a brain should be able to tie that knot."	Coercive power
"Well, this company's got to shape up. I've never seen such a bunch of out-of-shape firefighters. Either get to the workout room for your assigned times, or lose your shift privileges."	Coercive power
"I'm sorry to have to deny you your promotion. You failed to meet certain standards . . ."	Coercive power

"Guess what, Rick? The captain asked me to help him with the scheduling. He says it's one of his least favorite jobs. Can you believe that? I thought I was the only one who disliked it."

Identification power

"No, I don't want to change. The captain always wears *his* SCBA in this situation. If he does it, I'll do it."

Identification power

"Yeah, I admire him. He and I sort of operate on the same wavelength. Do you know that we had a lot of the same ideas for the design of the new firehouse?"

Identification power

"Have you ever seen anyone who knows as much as she does? And I thought it would be the pits to have a woman for captain."

Expert power

No, I think we should follow Marty's advice. After all, he is the one with the EMT training here."

Expert power

"Take that problem to Sam. He's always clever about knowing how to handle that type of thing."

Expert power

"The captain doesn't like the ruling any better than we do. He's only following departmental policy."

Legitimate power

"The mayor has requested that he ride the new engine in this parade."

Legitimate power

"The chief expects us to speak to each of the three grade schools during Fire Prevention Week."	Legitimate power

- Arrange for needed equipment and materials:
  ___ Overhead projector/screen
  ___ Transparencies
      Transparency 6-1—A Leader Wears Many Hats
      Transparency 6-2—Leadership, What Is It?
      Transparency 6-3—Leaders on Leadership
      Transparency 6-4—Theory X
      Transparency 6-5—Theory Y
      Transparency 6-6—Theory Z
      Transparency 6-7—Theories X, Y, and Z
      Transparency 6-8—Leadership Styles
      Transparency 6-9—Power
      Transparency 6-10—Power Types
      Transparency 6-11—An Effective Leader
      Transparency 6-12—Rise to New Heights of Leadership
  ___ Flipchart paper and easel
  ___ Prepared flipcharts
      Flipchart 6-1—Definition of Leadership
      Flipchart 6-2—Bureaucratic Leadership Style
      Flipchart 6-3—Single-Issue Leadership Style
      Flipchart 6-4—Middle-of-the-Road Leadership Style
      Flipchart 6-5—Dual-Issue Leadership Style
  ___ Masking tape for posting flipchart pages
  ___ **Student Guide**
      Activity Sheet 6-1—Analyze Leadership Styles
      Job Sheet 6-1—Organize and Lead a Group Session
  ___ Lesson Test
  ___ Other (list)

_____

_____

_____

REFERENCES CONSULTED

Leadership. Stillwater, Oklahoma: Fire Protection Publications, 1987.

Fire Department Company Officer, 2nd ed., Stillwater, Oklahoma: Fire Protection Publications, 1990.

LESSON 6—PRESENTATION

ADMINISTRATION/REVIEW 10 minutes

Take	roll.	
Return	Lesson 5 Test and critique in class as a review.	

INTRODUCTION/MOTIVATION 10 minutes

Display | Transparency 6-1. Discuss.

A LEADER WEARS MANY HATS
- MOTIVATOR
- PLANNER
- DECISION MAKER
- SUPERVISOR
- TEACHER
- EVALUATOR

Explain "In this lesson you will learn about the theories and styles of leadership, and you will examine the types of power that leaders assume. As an officer, you will wear the many hats of a leader, and you will influence your company through your leadership skills."

Refer students to Lesson 6 Study Objectives in the **Student Guide.** Discuss.

Ask What is leadership?"

> Guide student responses to the following definition: *Leadership is the ability to influence others.*

Display Transparency 6-2. Discuss why leadership is *not* a position one fills.

LEADERSHIP
WHAT IS IT? — ACHIEVEMENT OF ORGANIZATIONAL GOALS THROUGH OTHERS
WHAT IS IT NOT? — A POSITION ONE FILLS

Display Transparency 6-3.

Discuss these definitions of leadership as they compare to the definition in Transparency 6-2 and the definitions formulated earlier by the class. Write the consensus definition of leadership on Flipchart 6-1 and post.

> DEFINITION OF LEADERSHIP

LEADERSHIP THEORIES 15 minutes

Display and discuss Transparency 6-4.

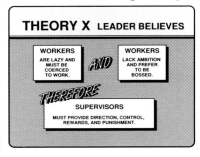

Display and discuss Transparency 6-5.

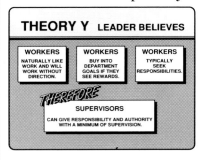

Display and discuss Transparency 6-6.

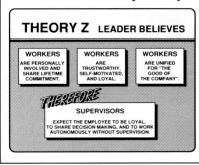

Summarize | subject block:

Display | Transparency 6-7.

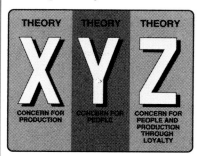

- Real-life concern for production and people can exist simultaneously. Theories X and Y fail except when directed toward specific individuals. Theory Z fails if the individuals being supervised cannot develop total commitment and loyalty to the company.
- When applied, management theory should not be all X, all Y, nor all Z. The company officer should select a mix of the three theories and a management and leadership style appropriate for the department and its individual members.

LEADERSHIP STYLES 30 minutes

Explain | that there are four basic leadership styles: bureaucratic, single-issue, middle-of-the-road, and dual-issue.

Discuss | bureaucratic leadership style by asking the following questions:

- "How does your textbook define bureaucratic leadership?"
- "What are some of the characteristics of this leadership style?"
- "What characteristics is the worker apt to exhibit under this type of leadership?"
- "What real-life examples of this type of leadership can you recall?"

Post | Flipchart 6-2 to summarize discussion.

> BUREAUCRATIC LEADERSHIP STYLE
>
> Strong social atmosphere
>
> Little stress-producing job-related competition
>
> Low concern for people
>
> Low concern for production
>
> Group pressure discourages overachievement
>
> Supervisor discourages individual initiative
>
> Workers can't meet job-oriented self-esteem needs

Discuss single-issue leadership by asking the following questions:

- "How does your textbook describe single-issue leadership?"
- "What are some of the characteristics of this leadership style?"
- "What characteristics is the worker apt to exhibit under this type of leadership?"
- "What real-life examples of this type of leadership can you recall?"

Post Flipchart 6-3 to summarize discussion.

> SINGLE-ISSUE LEADERSHIP STYLE
> Concern for *either* production *or* worker needs
> Production issue associated with Theory X
> Worker issue associated with Theory Y
> Worker's self- and social needs not met
> Moderately motivated workers quit
> High turnover rate and discontented workers

Discuss middle-of-the-road leadership style by asking the following questions:

- "How does your textbook describe middle-of-the-road leadership?"
- "What are some of the characteristics of this leadership style?"
- "What characteristics is the worker apt to exhibit under this type of leadership?"
- "What real-life examples of this type of leadership can you recall?"

Post Flipchart 6-4 to summarize discussion.

> MIDDLE-OF-THE-ROAD LEADERSHIP STYLE
> Moderate concern for production and workers
> Provides little direction
> Is way for supervisor to "get by"
> Leads to mediocrity
> Worker not motivated to show initiative
> Moderately motivated workers quit

Discuss dual-issue leadership style by asking the following questions:

- "How does your textbook describe dual-issue leadership?"

- "What are some of the characteristics of this leadership style?"
- "What characteristics is the worker apt to exhibit under this type of leadership?"
- "What real-life examples of this type of leadership can you recall?"

Post | Flipchart 6-5 to summarize discussion.

DUAL-ISSUE LEADERSHIP STYLE

High concern for production and worker

Workers can fulfill most of their self- and social needs

Workers share in decision making and show initiative

Much job satisfaction and little turnover

Summarize | subject block by displaying and reviewing Transparency 6-8 and by making the following points:

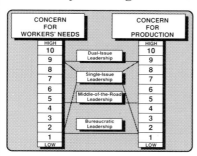

- The four basic leadership styles—bureaucratic, single-issue, middle-of-the-road, and dual-issue—place different degrees of emphasis on production and worker needs; each style has its strengths and weaknesses.

- Even in the same situation, leadership styles will (and should) vary from individual to individual. Therefore, factors beyond your control will determine which leadership style should be used.

- When selecting a leadership style, keep in mind 1) your personality, 2) the employee's personality, and 3) the situation.

- Know each of your subordinates: What are the individual's talents? Weaknesses? What motivates the individual? How mature is the individual? How dedicated?

TEACHING OPTION 6-1: ACTIVITY SHEET 6-1—ANALYZE LEADERSHIP STYLES
20 minutes

Refer | students to Activity Sheet 6-1 in **Student Guide**. Read aloud the case studies in Part I, and have students answer the questions as a group. Or break students into groups that complete the activity sheet and report to the class their consensus answers to the questions.

Assign | Activity Sheet 6-1, Part II for homework, due the next class session.

POWER STRUCTURES 15 minutes

Ask | "How would you define power?"

6-9

Summarize | student responses by displaying and reviewing Transparency 6-9.

A PERSON'S ABILITY TO INFLUENCE OTHERS

Display | Transparency 6-10. Discuss each of the types of power, stressing that the types of power presented are described from the employee's perception of the leader—not the leader's perception of his or her own power:

REWARD
COERCIVE
IDENTIFICATION
EXPERT
LEGITIMATE

Reward power—Employee's idea of leader's ability to grant rewards: promoting someone, giving a raise, expanding one's budget. Even stopping by an employee's desk to say "good morning" is a reward. Results depend on how the employee sees the leader's ability to provide an apparent reward.

Coercive power—Employee's observation of leader's ability to punish: a reprimand, scorn for mistakes made, denial of a raise or promotion, firing a person. The leader's withholding a promised or standard reward is also use of coercive power.

Identification power—Employee's impression of the similarity between himself or herself and the leader, or the desire to be like the leader. Many employees are willing to be influenced by leaders whom they feel are friends and can be trusted at a personal level.

Expert power—Employee's concept that a leader's expertise, intelligence, knowledge, or approach can help the employee adapt to a complex world. Members of society look for leaders who have a broad base of knowledge, or leaders who have the ability to apply their knowledge to solve problems.

Legitimate power—Employee's understanding that authority is given to the leader by a legitimate agent such as the fire department or city.

TEACHING OPTION 6-2: POWER PLAY 15 minutes

Divide | the class into two groups (teams) of equal numbers. If there is an odd number in the class, make the "odd" student the reader/scorekeeper. If not, you assume that duty.

Provide | the reader/scorekeeper with the Power Play flashcards prepared earlier. This person should also have chalk to keep score with, and should write "Team 1" and "Team 2" on the chalkboard.

Direct | the reader to read the situations on the cards while the teams compete with each other to see who can be first to identify the type of power described in each situation. Each time a team is first to identify the type of power used, it receives 10 points. The team with the most points after the last card is read, wins. If you wish, you may reward the members of this team by giving them full credit for Lesson 6 test question 4.

CHARACTERISTICS OF EFFECTIVE LEADERSHIP 20 minutes

Ask | students to think of presidents, politicians, club leaders, school leaders, military or fire service officers, and teachers. Ask the following questions and record student responses on the chalkboard.

- "As you have watched effective formal leaders, what characteristics have you noticed?"

- "What ineffective leadership traits have you seen on the job by those filling a formal position of leadership?"

Display Transparency 6-11. Compare the traits listed on the transparency to those presented by the class during your question session.

AN EFFECTIVE LEADER

PRODUCTIVITY

EFFECTIVENESS

- MAKES OTHER PEOPLE FEEL STRONG
- STRUCTURES COOPERATIVE RELATIONSHIPS
- RESOLVES CONFLICTS
- PROMOTES GOAL-ORIENTED THINKING AND BEHAVIOR
- BUILDS OTHER'S TRUST

Discuss each of the leadership traits. Ask students to provide job-related examples of each characteristic.

An effective leader. . .

Makes others feel strong—Employees are treated as equals and made to feel that they control their fate and can influence their future and their environment. When people feel strong, they enjoy their work, feel personally involved, and are motivated toward high production. Expert and identification power are most often used to make employees feel strong.

Structures cooperative rather than competitive relationships—The leader encourages cooperative relationships so that employees share company goals and cooperate with each other and other companies to achieve the goals of the department. A competitive relationship encourages reward power (winning) and coercive power (losing), whereas a cooperative relationship allows the officer to establish influence through legitimate and expert power.

Resolves conflicts—Leader handles conflict by confronting it with employees and resolving it according to established policies and procedures and in a cooperative manner. The leader does not avoid problems by denying they exist, by forcing a solution without consulting the individuals involved, or by smoothing over the problem. Conflict resolution comes from a leader's wise use of legitimate and expert power.

Promotes goal-oriented thinking and behavior—The leader uses team-building techniques and identification and legitimate power to orient group thinking and behavior toward achieving company and department goals. Team members depend on one another and work better together when united toward a common goal.

Builds other's trust—By building the employee's trust, the leader is put in a position of informal leadership, not just the formal leadership role by rank. This action often generates employee identification, which is one type of power.

SUMMARY: RISE TO NEW HEIGHTS OF LEADERSHIP 15 minutes

Summarize | lesson by explaining that company officers in their roles of formal group leaders have great influence on the company. Officers should study the different leadership styles, types of power, and leadership characteristics so they can select the most appropriate applications for particular situations.

Ask | individual students to briefly identify each of the following:

- Management Theory X, Theory Y, and Theory Z
- Bureaucratic, single-issue, middle-of-the-road, and dual-issue leadership styles
- Reward, coercive, identification, expert, and legitimate power
- Characteristics of an effective leader

Display | and review Transparency 6-12.

RISE TO NEW HEIGHTS OF LEADERSHIP

- Use of power
- Understanding human nature
- Motivation
- Concern for people
- Concern for production
- Getting work done
- Teamwork
- Trust
- Inspiration

Ask | for and answer final questions; remind students to prepare for the next class session by completing Part II of Activity Sheet 6-1, and by reading Chapter 6 in **Company Officer**.

LESSON TEST 30 minutes

Distribute | Lesson 6 Test and administer to evaluate the students' comprehension of the cognitive behaviors taught in this lesson.

Lesson 6

TRANSPARENCY SUMMARY

Lesson 6

Transparency 6-1

A LEADER WEARS MANY HATS

- MOTIVATOR
- PLANNER
- DECISION MAKER
- SUPERVISOR
- TEACHER
- EVALUATOR

Transparency 6-2

LEADERSHIP

WHAT IS IT?

ACHIEVEMENT OF ORGANIZATIONAL GOALS THROUGH OTHERS

WHAT IS IT NOT?

A POSITION ONE FILLS

Transparency 6-3

LEADERS ON LEADERSHIP

"The art of getting someone to do what you want done — because he wants to do it." — Dwight D. Eisenhower

"Leadership inspires people to exercise their best qualities." — Robert F. Kennedy

Transparency 6-4

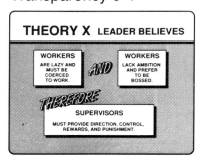

THEORY X LEADER BELIEVES

WORKERS ARE LAZY AND MUST BE COERCED TO WORK. *AND* WORKERS LACK AMBITION AND PREFER TO BE BOSSED.

THEREFORE

SUPERVISORS MUST PROVIDE DIRECTION, CONTROL, REWARDS, AND PUNISHMENT.

Transparency 6-5

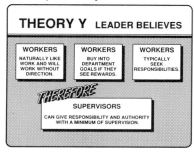

THEORY Y LEADER BELIEVES

WORKERS NATURALLY LIKE WORK AND WILL WORK WITHOUT DIRECTION. WORKERS BUY INTO DEPARTMENT GOALS IF THEY SEE REWARDS. WORKERS TYPICALLY SEEK RESPONSIBILITIES.

THEREFORE

SUPERVISORS CAN GIVE RESPONSIBILITY AND AUTHORITY WITH A MINIMUM OF SUPERVISION.

Transparency 6-6

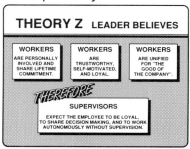

THEORY Z LEADER BELIEVES

WORKERS ARE PERSONALLY INVOLVED AND SHARE LIFETIME COMMITMENT. WORKERS ARE TRUSTWORTHY, SELF-MOTIVATED, AND LOYAL. WORKERS ARE UNIFIED FOR "THE GOOD OF THE COMPANY".

THEREFORE

SUPERVISORS EXPECT THE EMPLOYEE TO BE LOYAL, TO SHARE DECISION MAKING, AND TO WORK AUTONOMOUSLY WITHOUT SUPERVISION.

Transparency 6-7

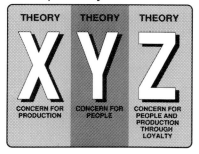

THEORY X CONCERN FOR PRODUCTION

THEORY Y CONCERN FOR PEOPLE

THEORY Z CONCERN FOR PEOPLE AND PRODUCTION THROUGH LOYALTY

Transparency 6-8

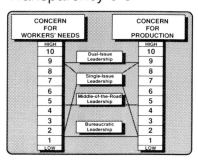

CONCERN FOR WORKERS' NEEDS | CONCERN FOR PRODUCTION

Dual-Issue Leadership
Single-Issue Leadership
Middle-of-the-Road Leadership
Bureaucratic Leadership

Transparency 6-9

POWER

A PERSON'S ABILITY TO INFLUENCE OTHERS

Transparency 6-10

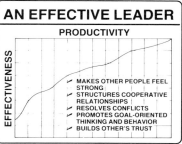

POWER

- REWARD
- COERCIVE
- IDENTIFICATION
- EXPERT
- LEGITIMATE

Transparency 6-11

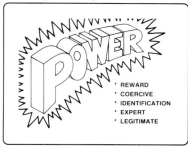

AN EFFECTIVE LEADER

PRODUCTIVITY

EFFECTIVENESS

- MAKES OTHER PEOPLE FEEL STRONG
- STRUCTURES COOPERATIVE RELATIONSHIPS
- RESOLVES CONFLICTS
- PROMOTES GOAL-ORIENTED THINKING AND BEHAVIOR
- BUILDS OTHER'S TRUST

Transparency 6-12

RISE TO NEW HEIGHTS OF LEADERSHIP

- Use of power
- Understanding human nature
- Motivation
- Concern for people
- Concern for production
- Getting work done
- Teamwork
- Trust
- Inspiration

FLIPCHART SUMMARY

Flipchart 6-1

DEFINITION OF LEADERSHIP

Flipchart 6-2

BUREAUCRATIC LEADERSHIP STYLE
Strong social atmosphere
Little stress-producing job-related competition
Low concern for people
Low concern for production
Group pressure discourages overachievement
Supervisor discourages individual initiative
Workers can't meet job-oriented self-esteem needs
Highly motivated workers quit

Flipchart 6-3

SINGLE-ISSUE LEADERSHIP STYLE
Concern for *either* production *or* worker needs
Production issue associated with Theory X
Worker issue associated with Theory Y
Worker's self- and social needs not met
Moderately motivated workers quit
High turnover rate and discontented workers

Flipchart 6-4

MIDDLE-OF-THE-ROAD LEADERSHIP STYLE
Moderate concern for production and workers
Provides little direction
Is a way for supervisor to "get by"
Leads to mediocrity
Worker not motivated to show initiative
Moderately motivated workers quit

Flipchart 6-5

DUAL-ISSUE LEADERSHIP STYLE
High concern for production and worker
Workers can fulfill most of their self-and social needs
Workers share in decision making and show initiative
Much job satisfaction and little turnover

LESSON 6 TEST

Name _____ **Date** _____

Note: Each test item has its own criterion standard. To show mastery of each tested objective, you must achieve a required number of points. The points you must achieve are listed first, followed by the points possible. For example, in a test item designated (9/12), you must achieve 9 of the 12 possible points.

---

1. Define *leadership*. (9/9).

   Leadership _____

   _____

2. Recognize the three theories of leadership. Match the leadership theories on the right with their characteristics on the left. Write the correct numbers in the blanks. Numbers will be used more than once. (2 pts. each, 24/30).

   _____a. Workers lack ambition.

   _____b. Supervisors can give responsibility and authority with a minimum of supervision.

   _____c. Workers seek responsibility.

   _____d. Workers are unified "for the good of the company."

   _____e. Workers are lazy and must be coerced to work.

   _____f. People are the most important.

   _____g. Workers naturally like to work and will work without direction.

   _____h. Supervisors expect the employee to share decision making and to work autonomously without supervision.

   _____i. Work and social life closely related.

   _____j. Production is the most important.

   _____k. Worker is self-directed.

   _____l. Supervisors must provide direction, control, rewards, and punishment.

   _____m. Worker seeks responsibility yet works cooperatively.

   _____n. Worker prefers direction and finds work distasteful.

   _____o. The group as a working family is most important.

   1. Theory X
   2. Theory Y
   3. Theory Z

---

3. Distinguish among leadership styles. Label the following leadership styles with their correct names. (4 pts. each, 16/16).

_____a. High concern for production and worker; workers can fulfill most of their self- and social needs; workers share in decision making and show initiative; much job satisfaction and little turnover.

_____b. Moderate concern for production and workers; provides little direction; is way for supervisor to "get by"; leads to mediocrity; worker can show initiative but is not motivated to do so; moderately motivated workers quit.

_____c. Concern for _either_ production _or_ worker needs; production issue associated with Theory X; worker issue associated with Theory Y; worker's self- and social needs not met; moderately motivated workers quit.

_____d. High turnover rate and discontented workers; strong social atmosphere; little stress-producing job-related competition; low concern for people; low concern for production; group pressure discourages overachievement; supervisor discourages individual initiative; workers can't meet job-oriented self-esteem needs; highly motivated workers quit.

4. Describe the types of power used by leaders. (4 pts. each, 12/20).

a. Reward power _____

_____

b. Coercive power _____

_____

c. Identification power _____

_____

d. Expert power _____

_____

e. Legitimate power _____

_____

5. List the five characteristics of an effective leader. (5 pts. each, 15/25).

a. _____

b. _____

c. _____

d. _____

e. _____

SCORING:	Objective Number	Points Needed/Possible	Points Achieved	Additional Study Needed	
				Yes	No
	1	9/9	_____	☐	☐
	2	24/30	_____	☐	☐
	3	16/16	_____	☐	☐
	4	12/20	_____	☐	☐
	5	15/25	_____	☐	☐
	TOTALS	76/100	_____	☐	☐

ANSWERS TO LESSON 6 TEST

1. Ability to influence others

2. a. 1
    b. 2
    c. 3
    d. 3
    e. 1
    f. 2
    g. 2
    h. 3
    i. 3
    j. 1
    k. 1
    l. 1
    m. 2
    n. 1
    o. 3

3. a. Dual-issue
    b. Middle-of-the-road
    c. Single-issue
    d. Bureaucratic

4. a. Leader's ability to grant rewards such as recognition, raises, praise
    b. Leader's ability to punish or to withhold rewards
    c. Employee's desire to be like the leader
    d. Leader's expertise, intelligence, knowledge, problem-solving ability
    e. Authoritative power given to the leader by the department

5. a. Makes others feel strong
    b. Structures cooperative relationships
    c. Resolves conflicts
    d. Promotes goal-oriented thinking and behavior
    e. Builds other's trust

STUDENT GUIDE

COMPANY OFFICER

For ifsta Company Officer

LESSON 6 STUDY OBJECTIVES

After completing Chapter 5 of **Company Officer** and related activities, you will be able to—

1. Define *leadership*.

2. Recognize the three theories of leadership.

3. Distinguish among leadership styles.

4. Describe the types of power used by leaders.

5. List the five characteristics of an effective leader.

6. Demonstrate the ability to organize and lead a group session.

STUDY SHEET

This study sheet is intended to help you learn the material in Chapter 5 of **Company Officer**. It is designed to be used for self-study, but may also be used to review material that will be covered in the lesson and content review test.

Chapter Vocabulary

Be sure you know the chapter-related meanings of the following terms:

- Leadership
- Inherent
- Autonomous
- Bureaucratic
- Coercive

Study Questions & Activities

1. List the basic principles of McGregor's Theory X and Theory Y leadership/management styles.

2. List the basic principles of the Theory Z leadership/management style.

3. List five things that motivate you to want to work.

4. Explain the characteristics of a bureaucratic leadership style.

5. Explain single-issue, dual-issue, and middle-of-the-road leadership styles.

6. What are the key characteristics of a leader?

7. Choose a leader whom you know well (teacher, officer, clergy, politician, etc.) and analyze that person's leadership style. Which style does this leader use primarily? Which of the key leadership characteristics does this person display? How does this individual motivate his or her employees or subordinates? Does this person seem to care more for production or people? Justify your answers with specific examples.

8. Explain and give examples of each of the following types of power:

   - Reward power
   - Coercive power
   - Identification power
   - Expert power
   - Legitimate power

Optional Assignment

Read chapters 1 through 4 in *Effective Supervisory Practices,* 3rd. ed. by International City Management Association, Washington D.C., 1989.

ACTIVITY SHEET 6-1
ANALYZE LEADERSHIP STYLES

Introduction

Most people lean more toward one leadership style than another, but it is important to remember that while the style used should be comfortable, it should not remain static. The style chosen must be appropriate for the situation and the employee, as well as for the leader. Part I of this activity sheet is designed to allow you to analyze the leadership styles of three fire service officers. In Part II, you will analyze the leadership styles of two captains.

Directions

Read the situations in Parts I and II carefully and then answer the questions that follow.

PART I Information

Chief Westhoff reviewed the monthly shift reports yesterday. He noticed that the less experienced firefighters on "A" and "B" shifts took a lot of time off on sick leave. He also noticed that over the past month, several rookie firefighters requested transfers to "C" shift. To accurately define the problem, Chief Westhoff called in the three shift commanders. He asked each for an opinion of the problem and the possible actions to take.

Captain Davis of "A" Shift explained that the quality of recruits was down, especially among the younger firefighters. He pointed out that most of the new recruits were lazy and went out of their way to avoid routine tasks such as housekeeping and station maintenance. Davis added, "Everyone knows that tradition has the probies working their way into the company. That's the way it's always been. When they fail to complete a routine task, I simply double the length of time they're posted to that duty."

Captain Carlson of "B" shift said that she had no complaints about the new recruits' attitudes, but their skill level was low. She named the training academy as part of the problem, citing the academy for not training the firefighters on a continuing basis. Carlson thought the recruits lacked confidence on the fireground and reported that they complained of inactivity between runs. "I have encouraged company members to bring reading materials with them to occupy their slack time. We have also started a weekly championship card game to keep their spirits up," she concluded.

Captain Walker of "C" shift reported that the firefighters on his shift seemed eager to learn and to work toward the company's goals. He reported that upon assignment, he paired each recruit with an older firefighter. He has the two firefighters perform the same tasks and duties for the first six months, and then the recruit is assigned to another mature firefighter for the remainder of the probationary year. This enriches the job of the mature firefighters by giving them training responsibility, which prepares them for promotion. This technique also gives the new firefighters job confidence and builds loyalty and team unity. Captain Walker also reported that he was working on increasing the new recruits' skill development through short training sessions and increased drill time. In addition, he encouraged the firefighters to interact socially, and close friendships had been formed.

PART I Activity

1. Which managerial theory does each captain most closely exhibit?

_____

_____

_____

2. Which concern (production or people) does each captain most prominently display? Why? Give specific examples to support your conclusions.

_____

_____

_____

_____

_____

_____

3. Which type of power does each captain most closely express in his or her comments? Why? Give specific examples to support your conclusions.

_____

_____

_____

_____

_____

_____

Part II Information

Captain Williams receives budget instructions from his chief, with deadlines for completion. He skims the instructions, noticing that his budget proposal is due in six weeks.

First he checks with other officers to see how much they plan to ask for their budget and how they plan to justify their requests. It appears to Williams that they are putting him off with comments such as "I won't know until I've studied my options," or "I need my team's recommendations before I make any final decisions."

With two weeks left, Williams is struggling with the budget, trying to show that he is an innovative leader who knows how to budget for his division. Alan enters and explains that since he knows it is budget time, he has some suggestions for word processing equipment that is needed. He asks if Williams would like to hear his suggestions. "Sure, sure. I'm always open. What is it?"

Alan outlines the problems with the present office set-up, and presents estimates on three systems with accompanying software. He recommends one of the three, based not only on price but also on maintenance costs and user needs.

Officer Williams replies, "That's good of you, Alan, but I'm going to do better than that. I'm requesting a Supersystem microcomputer that beats all others hollow. We need some sophisticated equipment to get the job done around here. Now don't you worry. I'll see that we get it."

One week before the budget is due, Williams is still dropping by the coffee machine to ask how the budget proposal is coming for the others. Their typical reply is "Going great" or "Challenging again this year, heh?"

Williams plans on dedicating this week before the budget is due to finalizing "what's been in his head all along." The weekend before finalizing the budget, he reads an article in the newspaper by a dissenting economist who predicts inflation will be 2 percent above anticipated figures. He decides he will use that percentage in his budget to figure inflationary costs on materials and supplies.

When he arrives at the station on Monday, he finds that one employee is ill and another has a sick child at home. He had forgotten he was scheduled to lead two grade school classes on a tour of the station and he is short staffed. The time he had planned for his budget proposal is lost to these unexpected demands.

In the end, his proposal requests two new positions, a microcomputer, inflation 2 percent higher than required to be used in the budget instructions, and out-of-state travel for his on-going training. His proposal arrives on the chief's desk four days late.

Captain Thomas receives the same budget instructions from his chief. He schedules some time that day to read the budget instructions. Once he understands what is expected in the process, he plans his approach and sets deadlines in order to meet the six-week deadline date.

Next, Thomas calls a staff meeting to explain the necessary budget instructions. He outlines the purpose of the budget process—to provide resources for members to do their jobs—and reminds the team that they are to keep the purpose of the fire department in mind as they make their budget requests. He defines his responsibility and their responsibility in the budget process and then sets deadlines with them. He asks for and answers any questions.

Thomas then explains that he will take their request lists and summarize them for the whole group, adding any other requests that should be included. He will hold another staff meeting to clarify and prioritize the requests by staff consensus, based on the needs of the total division.

After a week, Thomas checks progress with each employee. Williams moans about not having enough time with two other special projects going, and Thomas assigns another employee to help Williams with one of his projects.

Sandy vaguely mumbles something about not seeing any reason why she should do the boss's work. Thomas reminds her that she has been in need of some supplies they both know are badly needed. "This is your opportunity to see that we get the supplies, but it will take a thorough justification of need. I see no reason why it can't be a part of the final budget request." With renewed interest, each employee has his or her report ready by the deadline.

Captain Thomas prepares and distributes the anticipated summary report to all employees and schedules a staff meeting three days later to give everyone a chance to review the summary. At the staff meeting, he asks each person to briefly describe his or her request and to answer any questions the rest of the staff have. Then the group prioritizes all requests on a consensus basis.

Once this is completed, Captain Thomas has two weeks to compile the budget proposal for his division in accordance with instructions. Officer Thomas's proposal is on his boss's desk one day early, with a copy in each employee's in-basket.

Part II Activity

1. Which of the two captains, Williams or Thomas, displayed the most effective leadership style?

   _____

   _____

2. What leadership characteristics are shown by Thomas?

   _____

   _____

3. What are the results of Thomas's leadership approach in this situation?

   _____

   _____

4. What characteristics and attitudes keep Williams from being an effective leader?

   _____

   _____

5. What are the results of Williams' leadership approach in this situation?

   _____

   _____

ANSWERS TO ACTIVITY SHEET 6-1

Part I

1. Captain Davis—Theory X
   Captain Carlson—Theory Y
   Captain Walker—Theory Z

2. Captain Davis—Production
   Captain Carlson—People
   Captain Walker—People and Production

3. Captain Davis—Coercive
   Captain Carlson—Reward
   Captain Walker—Identification, Expert, Legitimate

Part II

1. Thomas

2. Ability to get things done through others; ability to help individuals make the organizational goals their own; ability to plan, organize, schedule, supervise, set priorities, take responsibility, train employees, build a team; ability to make others feel good about themselves; thorough and attentive to details while keeping the total picture in mind.

3. At a minimum, the results include a sound budget proposal, continued high morale, and responsible, well-trained employees who are familiar with the budget process.

4. Does not know how to get things done through others; does not communicate well with employees or peers; is motivated by self-interest; does not consult instructions thoroughly and does not plan for process; is patronizing to employees and "milks" peers for their ideas on what he should be doing; is not well disciplined, and therefore is tardy with the proposal; does not assess needs for budget resources realistically, and is not prepared for the unforeseen.

5. At a minimum, the results would include an unsound budget proposal, low morale, and employees who are not familiar with nor trained in the budget process.

JOB SHEET 6-1
ORGANIZE AND LEAD A GROUP SESSION

Prerequisites: **Company Officer**, Chapter 5; Activity Sheet 6-1

Introduction: A guided group session is designed by the leader to develop group understanding, or perhaps to bring the group to general agreement through talk and reflective thinking. Its aims are to 1) stimulate thought and analysis, 2) encourage interpretations of the facts, and 3) develop new attitudes or change old ones. With good leadership, evidence on a crucial issue or problem is brought out, evaluated, and some conclusions are reached.

Equipment and Materials: Classroom
Chalk
Chalkboard
Group(s) of 4 to 8 people

Operations	Key Points
Plan for group session: 1. Determine a topic of concern and interest to the group.	1. You will have to guide the topic selection, but the group should feel that they participated in the process and that the topic is relevant to their real needs and interests.
2. Outline direction session is going to take.	2. What must be accomplished at the session? Must the group arrive at a conclusion? Suggest action to be taken? Break into committees?
3. Prepare leading questions to be asked during the session.	3. Write down these questions and any suggested answers that occur to you. Be prepared with current information.
4. Prepare the physical setting by arranging the chairs in a semicircle, in clusters, or around a large table, locating yourself near the apex of the group or wherever you can be easily seen and heard.	4. Plan for a close but informal setting.
Lead group session: 5. Introduce topic of session, the general limits of the topic, and the time schedule agreed upon.	5. Try not to simply announce the topic. A good way to begin is to introduce the topic and scope of the session in question form: "What is the scope of the problem of *Aids* in the fire service today, and what actions can be taken to avoid infection?"

Operations	Key Points
6. Explain the purpose of the session, making the problem clear by stating it in specific and direct terms.	
7. Ask leading questions to initiate discussion among group members.	7. Examples: How serious is the problem of AIDS infection on a national level? Is infection a problem in this department? What can the department do about the problem? What can the company do?
8. Keep the discussion on the topic; guide the group, but do not take a position or monopolize.	8. Lead by clarifying, defining, providing facts, and helping the group to organize or to understand departmental policy or procedure.
9. Establish an atmosphere of friendly cooperation in which all group members can interact and participate.	9. Group members will find an adversary approach or one of aggressive competition threatening.
10. Call on those who show nonverbal interest, but who do not volunteer.	10. The skillful leader develops an awareness of facial expressions and is sensitive to the enthusiasm and attitudes of the group.
11. Summarize the discussion periodically when needed.	11. "So, if I hear you right, you are saying . . . "
12. Play the "devil's advocate" as appropriate to get the group to look at the other isde of a question.	12. This kind of openness requires a leader who is free from a drive to dominate, who is perfectly secure, who is willing to sometimes be a follower, and who can restrain his or her own desire to talk.
Close the session: 13. Close the session by summarizing the group's conclusions or decisions.	13. Help the group come to some conclusions or a consensus opinion.
14. Suggest courses of action or ways of using insights gained from the session.	

JS 6-1

JS 6-1

PERFORMANCE EVALUATION

Name _____ Date _____

Evaluator _____ Overall Competency Rating _____

Rating Scale:

4—Skilled—Can perform independently with no additional training

3—Moderately Skilled—Has performed independently during the training session; limited additional training may be required

2—Limited Practice—Has practiced during training session; additional training required

1—Exposure Only—General information provided with no practice time or supervision; additional training required

0—No Exposure—No information nor practice provided during training program; complete training required

N/A—Not applicable

Evaluator's Note: Find the overall competency rating by averaging the performance level ratings and rounding to a whole number. Record above and on the Competency Profile.

Level of Performance

1　2　3　4

Planning the session—

1. Assisted group in determining topic of concern and interest. ☐ ☐ ☐ ☐

2. Outlined direction session was going to take. ☐ ☐ ☐ ☐

3. Prepared leading questions to be asked during session. ☐ ☐ ☐ ☐

4. Prepared the physical setting for the session so that:

　a. all could see and hear. ☐ ☐ ☐ ☐

　b. an informal and comfortable environment was provided. ☐ ☐ ☐ ☐

Leading the session—

5. Introduced the topic. ☐ ☐ ☐ ☐

6. Explained the purpose of the session. ☐ ☐ ☐ ☐

7. Asked leading questions. ☐ ☐ ☐ ☐

8. Kept the discussion on the topic. ☐ ☐ ☐ ☐

9. Ensured that participation was fairly balanced. ☐ ☐ ☐ ☐

10. Called on those who showed nonverbal interest, but who did not volunteer. ☐ ☐ ☐ ☐

Level of Performance

	1	2	3	4

11. Refrained from taking a position or monopolizing the group. ☐ ☐ ☐ ☐

12. Summarized the discussion periodically when needed. ☐ ☐ ☐ ☐

Closing the session—

13. Brought the session to a satisfactory conclusion. ☐ ☐ ☐ ☐

14. Suggested courses of action or ways of using the insights gained from the session. ☐ ☐ ☐ ☐

Competency Rating

Totals __ + __ + __ + __ = ___ = _____

15

Appendix B

LOCALLY CONSTRUCTED TRAINING AIDS

Locally constructed training aids can be valuable additions to a training program, especially for fire departments with limited financial resources. Such training aids as manikins and forcible entry props can be easily constructed using readily available resources. This appendix describes training aids that can easily be made and used to help teach manipulative skills.

There are certain considerations a fire department should take into account when using training aids. Although training aids cannot be the "real thing," they should be able to give students a realistic idea of the skill being taught. For example, if a forcible entry prop will open with just a slight push, it is not teaching firefighters forcible entry skills.

The training aids shown in this section are only examples of ways to use equipment to improve learning in your training program.

HOSE "A" RESCUE MANIKIN (Figure B.1)

1. Straight roll 10 (3 m) feet of 3-inch (77 mm) hose and secure with bolt to form head.

(**NOTE:** Use ¼ x 4-inch [6 mm by 100 mm] stove bolts with nuts on inside.)

2. Drill hose at center of 50-foot (15 m) section of 3-inch (77 mm) hose and attach to head with bolt.

To form torso, fold half of 50-foot (15 m) section about 8 inches (200 mm) wide (shoulder) and about 33 inches (840 mm) long (abdomen) and 8 inches (200 mm) (hip) from outside toward center. Accordion fold remaining hose inside. Fold the other half in the same manner. Secure folds by bolting flakes together starting at the center. (Figure B.2)

3. Attach legs by inserting the center of 17 feet (5.2 m) of 3-inch (77 mm) hose through the bottom loop of torso and fold remaining hose about 32 inches (815 mm) long and toward inside and bolt together.

4. Fold 12 feet (3.7 mm) of 2 ½-inch (65 mm) hose at center and attach to shoulder area with bolt. Fold remaining hose about 26 inches (660 mm) long and toward inside and bolt together for arm.

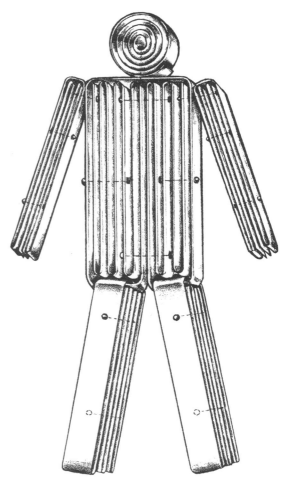

Figure B.1 An example of a Hose "A" Rescue Manikin.

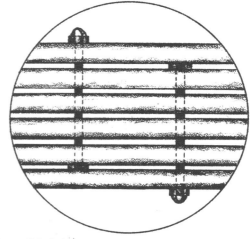

Figure B.2 Bolts secure the folds of the hose.

5. Add gloves, boots, coveralls, sweatshirt hood, or ski mask.

Courtesy of W.C. (Bill) Hulsey, Broken Arrow (Oklahoma) Fire Department, Training Division.

LENS COVER FOR SELF-CONTAINED BREATHING APPARATUS TRAINING (Figure B.3)

Cut 11x14-inch (280 mm by 350 mm) oval from blue denim or other dark material. Turn under ¼-inch (6 mm) seam and stitch. Stitch 18 inches (450 mm) of ¼-inch (6 mm) wide elastic around edge using zig zag stitch and stretching elastic as it is sewn.

Stretch lens cover over mask to simulate dark, smoky conditions while training with self-contained breathing apparatus.

Courtesy of W.C. (Bill) Hulsey, Broken Arrow (Oklahoma) Fire Department, Training Division.

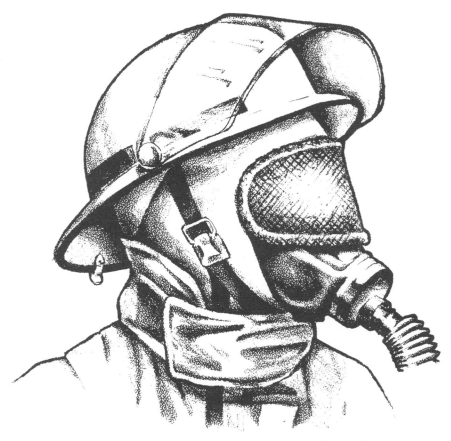

Figure B.3 Dark material over the lens cover simulates dark, smoky conditions.

FORCIBLE ENTRY PROP NO. 1 (Figure B.4)

A prehung door and exterior frame with supports to hold it upright makes a good forcible entry training prop. The frame gives, allowing firefighters to spring the door away from the frame for the bolt to pass the keeper. (Figure B.5)

Firefighters can practice forcible entry by prying with different tools and removing hinge pins. (Figure B.6)

Courtesy of W.C. (Bill) Hulsey, Broken Arrow (Oklahoma) Fire Department, Training Division.

Figure B.4 A prehung door makes a good forcible entry prop.

Figure B.5 The door can be sprung without damaging the frame.

Figure B.6 The hinge pins can be removed with various tools.

FORCIBLE ENTRY PROP NO. 2 (Figure B.7)

To practice breaking glass, a single window sash hung on a structure with pins or nails makes a good training prop. Firefighters can practice forcible entry by breaking the glass using different tools.

Courtesy of W.C. (Bill) Hulsey, Broken Arrow (Oklahoma) Fire Department.

Figure B.7 A single window sash makes a good training prop to simulate forcible entry by breaking a window.

FORCIBLE ENTRY PROP NO. 3 (Figure B.8)

Wooden pallets can be used to practice ventilation procedures. Lay flat or elevate.

Firefighters can use the training prop to simulate cutting roof members.

Courtesy of W.C. (Bill) Hulsey, Broken Arrow (Oklahoma) Fire Department, Training Division.

Figure B.8 Wooden pallets simulate cutting roof members.

VENTILATION PROP FOR A ROOF AREA (Figure B.9)

A portable roof can be placed on a flat-top roof so firefighters can practice ventilation and using a roof ladder.

Build it from scrap 2 x 4-foot (50 mm by 100 mm) lumber and cover it with plywood. The 2 x 4-foot (50 mm by 100 mm) frame can be bolted together, making the portable roof collapsible.

Firefighters can use the prop to simulate ventilation practices using different tools.

Courtesy of W.C. (Bill) Hulsey, Broken Arrow (Oklahoma) Fire Department, Training Division.

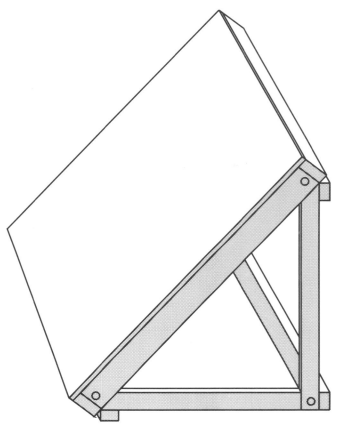

Figure B.9 This prop can be used to simulate roof ventilation.

HOSE BED FOR HOSE LOAD TRAINING (Figure B.10)

Materials:

One — 48 x 96 x ³/₄-inch (1 200 mm by 2 400 mm by 20 mm) plywood (bottom)

Two — 12 x 96 x ³/₄-inch (300 mm by 2 400 mm by 20 mm) plywood (sides)

One — 11 ¹/₄ x 95 ¹/₄ x ³/₄-inch (285 mm by 2 380 mm by 20 mm) plywood (inside divider)

One — 11 ¹/₄ x 48 x ³/₄-inch (285 mm by 1 220 mm by 20 mm) plywood (back)

Four — Shelf brackets (inside support)

Firefighters can use this prop to replace an actual hose bed. The advantages of this hose bed are that it does not take fire equipment out of service and it is low to the ground.

Courtesy of W.C. (Bill) Hulsey, Broken Arrow (Oklahoma) Fire Department, Training Division.

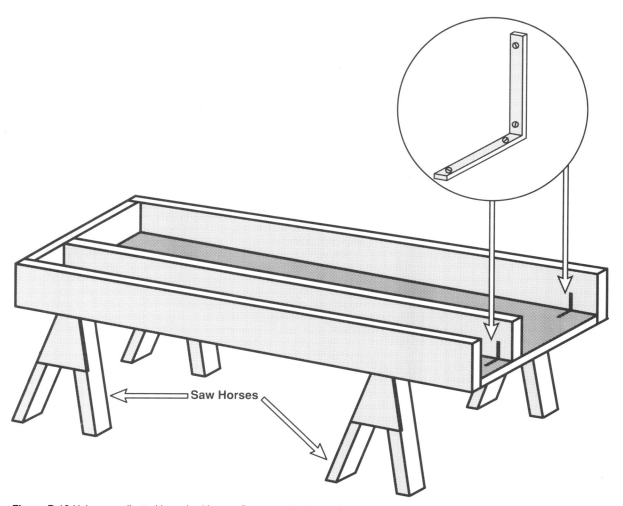

Saw Horses

Figure B.10 Using a replicated hose bed leaves fire apparatus in service.

FORCIBLE ENTRY — "TRY BEFORE YOU PRY" (Figure B.11)

Simple, but effective, forcible entry props can be constructed and reused at very little cost. The McCormack triangle, built and used at the Massachusetts Firefighting Academy, is one such prop.

The prop can be built from scrap lumber. A few 2 x 8-inch (50 mm by 200 mm) pieces, covered with plywood, form the floor. The three walls can be covered with a variety of common building materials. A door and one or two windows are incorporated. The roof covering can be any type common to your area. The roof covering is built over 2 x 4-inch (50 mm by 100 mm) or 2 x 6-inch (50 mm by 150 mm) material. All of the materials used should be easy to replace. This allows for continuous use of the prop. **Remember** — this is a forcible entry prop, not a burn building. Pipe or heavy channel iron can be used as a frame. The roof, walls, and floor panels are bolted on to the frame. Firefighters can use the prop to practice forcible entry by using different tools.

Credit: International Society of Fire Service Instructors "Instruct-O-Gram".

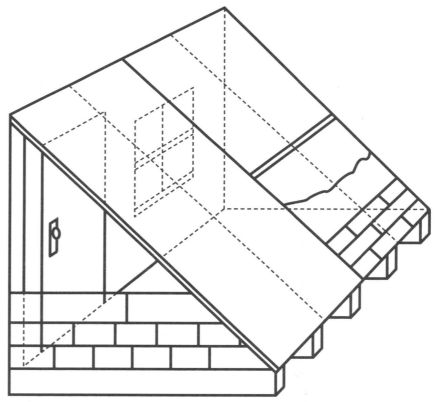

Figure B.11 The McCormack triangle is a simple and effective forcible entry prop.

Index

IFSTA MANUALS AND FPP PRODUCTS

**For a current catalog describing these and other products, call or write your local IFSTA distributor or Fire Protection Publications, IFSTA Headquarters, Oklahoma State University, Stillwater, OK 74078-0118.
Phone: 1-800-654-4055**

FIRE DEPARTMENT AERIAL APPARATUS
includes information on the driver/operator's qualifications; vehicle operation; types of aerial apparatus; positioning, stabilizing, and operating aerial devices; tactics for aerial devices; and maintaining, testing, and purchasing aerial apparatus. Detailed appendices describe specific manufacturers' aerial devices. 1st Edition (1991), 386 pages, addresses NFPA 1002.

STUDY GUIDE FOR AERIAL APPARATUS
The companion study guide in question and answer format. 1991, 140 pages.

AIRCRAFT RESCUE AND FIRE FIGHTING
comprehensively covers commercial, military, and general aviation. All the information you need is in one place. Subjects covered include: personal protective equipment, apparatus and equipment, extinguishing agents, engines and systems, fire fighting procedures, hazardous materials, and fire prevention. Over 240 photographs and two-color illustrations. It also contains a glossary and review questions with answers. 3rd Edition (1992), 247 pages, addresses NFPA 1003.

BUILDING CONSTRUCTION RELATED TO THE FIRE SERVICE
helps firefighters become aware of the many construction designs and features of buildings found in a typical first alarm district and how these designs serve or hinder the suppression effort. Subjects include construction principles, assemblies and their resistance to fire, building services, door and window assemblies, and special types of structures. 1st Edition (1986), 166 pages, addresses NFPA 1001 and NFPA 1031, levels I & II.

CHIEF OFFICER
lists, explains, and illustrates the skills necessary to plan and maintain an efficient and cost-effective fire department. The combination of an ever-increasing fire problem, spiraling personnel and equipment costs, and the development of new technologies and methods for decision making requires far more than expertise in fire suppression. Today's chief officer must possess the ability to plan and administrate as well as have political expertise. 1st Edition (1985), 211 pages, addresses NFPA 1021, level VI.

SELF-INSTRUCTION FOR CHIEF OFFICER
The companion study guide in question and answer format. 1986, 142 pages.

FIRE DEPARTMENT COMPANY OFFICER
focuses on the basic principles of fire department organization, working relationships, and personnel management. For the firefighter aspiring to become a company officer, or a company officer wishing to improve management skills, this manual helps develop and improve the necessary traits to effectively manage the fire company. 2nd Edition (1990), 278 pages, addresses NFPA 1021, levels I, II, & III.

COMPANY OFFICER STUDY GUIDE
The companion study guide in question and answer format. Includes problem applications and case studies. 1991, 243 pages.

ESSENTIALS OF FIRE FIGHTING
is the "bible" on basic firefighter skills and is used throughout the world. The easy-to-read format is enhanced by 1,600 photographs and illustrations. Step-by-step instructions are provided for many fire fighting tasks. Topics covered include: personal protective equipment, building construction, firefighter safety, fire behavior, portable extinguishers, SCBA, ropes and knots, rescue, forcible entry, ventilation, communications, water supplies, fire streams, hose, fire cause determination, public fire education and prevention, fire suppression techniques, ladders, salvage and overhaul, and automatic sprinkler systems. 3rd Edition (1992), 590 pages, addresses NFPA 1001.

STUDY GUIDE FOR 3rd EDITION OF ESSENTIALS OF FIRE FIGHTING
The companion learning tool for the new 3rd edition of the manual. It contains questions and answers to help you learn the important information in the book. 1992, 322 pages.

PRINCIPLES OF EXTRICATION
leads you step-by-step through the procedures for disentangling victims from cars, buses, trains, farm equipment, and industrial situations. Fully illustrated with color diagrams and more than 500 photographs. It includes rescue company organization, protective clothing, and evaluating resources. Review questions with answers at the end of each chapter. 1st Edition (1990), 365 pages.

FIRE CAUSE DETERMINATION
gives you the information necessary to make on-scene fire cause determinations. You will know when to call for a trained investigator, and you will be able to help the investigator. It includes a profile of firesetters, finding origin and cause, documenting evidence, interviewing witnesses, and courtroom demeanor. 1st Edition (1982), 159 pages, addresses NFPA 1021, Fire Officer I, and NFPA 1031, levels I & II.

FIRE SERVICE FIRST RESPONDER
provides the information needed to evaluate and treat patients with serious injuries or illnesses. It familiarizes the reader with a wide variety of medical equipment and supplies. **First Responder** applies to safety, security, fire brigade, and law enforcement personnel, as well as fire service personnel, who are required to administer emergency medical care. 1st Edition (1987), 340 pages, addresses NFPA 1001, levels I & II, and DOT First Responder.

FORCIBLE ENTRY
reflects the growing concern for the reduction of property damage as well as firefighter safety. This comprehensive manual contains technical information about forcible entry tactics, tools, and methods, as well as door, window, and wall construction. Tactics discuss the degree of danger to the structure and leaving the building secure after entry. Includes a section on locks and through-the-lock entry. Review questions and answers at the end of each chapter. 7th Edition (1987), 270 pages, helpful for NFPA 1001.

GROUND COVER FIRE FIGHTING PRACTICES

explains the dramatic difference between structural fire fighting and wildland fire fighting. Ground cover fires include fires in weeds, grass, field crops, and brush. It discusses the apparatus, equipment, and extinguishing agents used to combat wildland fires. Outdoor fire behavior and how fuels, weather, and topography affect fire spread are explained. The text also covers personnel safety, management, and suppression methods. It contains a glossary, sample fire operation plan, fire control organization system, fire origin and cause determination, and water expansion pump systems. 2nd Edition (1982), 152 pages.

FIRE SERVICE GROUND LADDER PRACTICES

is a "how to" manual for learning how to handle, raise, and climb ground ladders; it also details maintenance and service testing. Basic information is presented with a variety of methods that allow the readers to select the best method for their locale. The chapter on Special Uses includes: ladders as a stretcher, a slide, a float drag, a water chute, and more. The manual contains a glossary, review questions and answers, and a sample testing and repair form. 8th Edition (1984), 388 pages, addresses NFPA 1001.

HAZARDOUS MATERIALS FOR FIRST RESPONDERS

provides basic information on hazardous materials with sections on site management and decontamination. It includes a description of various types of materials, their characteristics, and containers. The manual covers the effects of weather, topography, and environment on the behavior of hazardous materials and control efforts. Pre-incident planning and post-incident analysis are covered. 1st Edition (1988), 357 pages, addresses NFPA 472, 29 CFR 1910.120 and NFPA 1001.

STUDY GUIDE FOR HAZARDOUS MATERIALS FOR FIRST RESPONDERS

The companion study guide in question and answer format also includes case studies that simulate incidents. 1989, 208 pages.

HAZARDOUS MATERIALS: MANAGING THE INCIDENT

takes you beyond the basic information found in **Hazardous Materials for First Responders**. Directed to the leader/commander, this manual sets forth basic practices clearly and comprehensively. Charts, tables, and checklists guide you through the organization and planning stages to decontamination. This text, along with the accompanying workbook and instructor's guide, provides a comprehensive learning package. 1st Edition (1988), 206 pages, helpful for NFPA 1021.

STUDENT WORKBOOK FOR HAZARDOUS MATERIALS: MANAGING THE INCIDENT

provides questions and answers to enhance the student's comprehension and retention. 1988, 176 pages.

INSTRUCTOR'S GUIDE FOR HAZARDOUS MATERIALS: MANAGING THE INCIDENT

provides lessons based on each chapter, adult learning tips, and appendices of references and suggested audio visuals. 1988, 142 pages.

HAZ MAT RESPONSE TEAM LEAK AND SPILL GUIDE

contains articles by Michael Hildebrand reprinted from *Speaking of Fire's* popular Hazardous Materials Nuts and Bolts series. Two additional articles from *Speaking of Fire* and the hazardous material incident SOP from the Chicago Fire Department are also included. 1st Edition (1984), 57 pages.

EMERGENCY OPERATIONS IN HIGH-RACK STORAGE

is a concise summary of emergency operations in the high-rack storage area of a warehouse. It explains how to develop a pre-emergency plan, what equipment will be necessary to implement the plan, type and amount of training personnel will need to handle an emergency, and interfacing with various agencies. Includes consideration questions, points not to be overlooked, and trial scenarios. 1st Edition (1981), 97 pages.

HOSE PRACTICES

reflects the latest information on modern fire hose and couplings. It is the most comprehensive single source about hose and its use. The manual details basic methods of handling hose, including large diameter hose. It is fully illustrated with photographs showing loads, evolutions, and techniques. This complete and practical book explains the national standards for hose and couplings. 7th Edition (1988), 245 pages, addresses NFPA 1001.

FIRE PROTECTION HYDRAULICS AND WATER SUPPLY ANALYSIS

covers the quantity and pressure of water needed to provide adequate fire protection, the ability of existing water supply systems to provide fire protection, the adequacy of a water supply for a sprinkler system, and alternatives for deficient water supply systems. 1st Edition (1990), 340 pages.

INCIDENT COMMAND SYSTEM (ICS)

was developed by a multiagency task force. Using this system, fire, police, and other government groups can operate together effectively under a single command. The system is modular and can be used to meet the requirements of both day-to-day and large-incident operations. It is the approved basic command system taught at the National Fire Academy. 1st Edition (1983), 220 pages, helpful for NFPA 1021.

INDUSTRIAL FIRE PROTECTION

is designed for the person charged with the responsibility of developing, implementing, and coordinating fire protection. A "must read" for fire service personnel who will coordinate with industry/business for pre-incident planning. The text includes guidelines for establishing a company policy, organization and planning for the emergency, establishing a fire prevention plan, incipient fire fighting tactics, an overview of interior structural fire fighting, and fixed fire fighting systems. 1st Edition (1982), 207 pages, written for 29 CFR 1910, Subpart L, and helpful for NFPA 1021 and NFPA 1031.

FIRE INSPECTION AND CODE ENFORCEMENT

provides a comprehensive, state-of-the-art reference and training manual for both uniformed and civilian inspectors. It is a comprehensive guide to the principles and techniques of inspection. Text includes information on how fire travels, electrical hazards, and fire resistance requirements. It covers storage, handling, and use of hazardous materials; fire protection systems; and building construction for fire and life safety. 5th Edition (1987), 316 pages, addresses NFPA 1001 and NFPA 1031, levels I & II.

STUDY GUIDE FOR FIRE INSPECTION AND CODE ENFORCEMENT

The companion study guide in question and answer format with case studies. 1989, 272 pages.

FIRE SERVICE INSTRUCTOR

explains the characteristics of a good instructor, shows you how to determine training requirements, and teach to the level of your class. It discusses the types, principles, and procedures of teaching and learning, and covers the use of effective training aids and devices. The purpose and principles of testing as well as test construction are covered. Included are chapters on safety, legal considerations, and computers. 5th Edition (1990), 326 pages, addresses NFPA 1041, levels I & II.

LEADERSHIP IN THE FIRE SERVICE

was created from the series of lectures given by Robert F. Hamm to assist in leadership development. It provides the foundation for getting along with others, explains how to gain the confidence of your personnel, and covers what is expected of an officer. Included is information on supervision, evaluations, delegating, and teaching. Some of the topics include: the successful leader today, a look into the past may reveal the future, and self-analysis for officers. 1st Edition (1967), 132 pages.

FIRE SERVICE ORIENTATION AND TERMINOLOGY

Fire Service Orientation and Indoctrination has been revised. It has a new name and a new look. Keeping the best of the old — traditions, history, and organization — this new manual provides a complete dictionary of fire service terms. To be used in conjunction with **Essentials of Fire Fighting** and the other IFSTA manuals. 3rd Edition (1993), addresses NFPA 1001.

PRIVATE FIRE PROTECTION AND DETECTION

provides a means by which fires may be prevented or attacked in their incipient phase and/or controlled until the fire brigade or public fire protection can arrive. This second edition covers information on automatic sprinkler systems, hose standpipe systems, fixed fire pump installations, portable fire extinguishers, fixed special agent extinguishing systems, and fire alarm and detection systems. Information on the design, operation, maintenance, and inspection of these systems and equipment is provided. 2nd Edition (1994).

PUBLIC FIRE EDUCATION

provides valuable information for ending public apathy and ignorance about fire. This manual gives you the knowledge to plan and implement fire prevention campaigns. It shows you how to tailor the individual programs to your audience as well as the time of year or specific problems. It includes working with the media, resource exchange, and smoke detectors. 1st Edition (1979), 169 pages, helpful for NFPA 1021 and 1031.

FIRE DEPARTMENT PUMPING APPARATUS

is the Driver/Operator's encyclopedia on operating fire pumps and pumping apparatus. It covers pumpers, tankers (tenders), brush apparatus, and aerials with pumps. This comprehensive volume explains safe driving techniques, getting maximum efficiency from the pump, and basic water supply. It includes specification writing, apparatus testing, and extensive appendices of pump manufacturers. 7th Edition (1989), 374 pages, addresses NFPA 1002.

STUDY GUIDE FOR PUMPING APPARATUS

The companion study guide in question and answer format. 1990, 100 pages.

FIRE SERVICE RESCUE PRACTICES

is a comprehensive training text for firefighters and fire brigade members that expands proficiency in moving and removing victims from hazardous situations. This extensively illustrated manual includes rescuer safety, effects of rescue work on victims, rescue from hazardous atmospheres, trenching, and outdoor searches. 5th Edition (1981), 262 pages, addresses NFPA 1001.

RESIDENTIAL SPRINKLERS A PRIMER

outlines United States residential fire experience, system components, engineering requirements, and issues concerning automatic and fixed residential sprinkler systems. Written by Gary Courtney and Scott Kerwood and reprinted from *Speaking of Fire*. An excellent reference source for any fire service library and an excellent supplement to **Private Fire Protection.** 1st Edition (1986), 16 pages.

FIRE DEPARTMENT OCCUPATIONAL SAFETY

addresses the basic responsibilities and qualifications for a safety officer and the minimum requirements and procedures for a safety and health program. Included in this manual is an overview of establishing and implementing a safety program, physical fitness and health considerations, safety in training, fire station safety, tool and equipment safety and maintenance, personal protective equipment, en- route hazards and response, emergency scene safety, and special hazards. 2nd Edition (1991), 366 pages, addresses NFPA 1500, 1501.

SALVAGE AND OVERHAUL

covers planning salvage operations, equipment selection and care, as well as describing methods and techniques for using salvage equipment to minimize fire damage caused by water, smoke, heat, and debris. The overhaul section includes methods for finding hidden fire, protection of fire cause evidence, safety during overhaul operations, and restoration of property and fire protection systems after a fire. 7th Edition (1985), 225 pages, addresses NFPA 1001.

SELF-CONTAINED BREATHING APPARATUS

contains all the basics of SCBA use, care, testing, and operation. Special attention is given to safety and training. The chapter on Emergency Conditions Breathing has been completely revised to incorporate safer emergency methods that can be used with newer models of SCBA. Also included are appendices describing regulatory agencies and donning and doffing procedures for nine types of SCBA. The manual has been thoroughly updated to cover NFPA, OSHA, ANSI, and NIOSH regulations and standards as they pertain to SCBA. 2nd Edition (1991), 360 pages, addresses NFPA 1001.

STUDY GUIDE FOR SELF-CONTAINED BREATHING APPARATUS

The companion study guide in question and answer format. 1991, 131 pages.

FIRE STREAM PRACTICES

brings you an all new approach to calculating friction loss. This carefully written text covers the physics of fire and water; the characteristics, requirements, and principles of good streams; and fire fighting foams. **Streams** includes formulas for the application of fire fighting hydraulics, as well as actions and reactions created by applying streams under a variety of circumstances. The friction loss equations and answers are included, and review questions are located at the end of each chapter. 7th Edition (1989), 464 pages, addresses NFPA 1001 and NFPA 1002.

GASOLINE TANK TRUCK EMERGENCIES

provides emergency response personnel with background information, general procedures, and response guidelines to be followed when responding to and operating at incidents involving MC-306/DOT 406 cargo tank trucks. Specific topics include: incident management procedures, site safety considerations, methods of product transfer, and vehicle uprighting considerations. 1st Edition (1992), 51 pages, addresses NFPA 472.

FIRE SERVICE VENTILATION

presents the principles and practices of ventilation. The manual describes and illustrates the safe operations related to ventilation, products of combustion, elements and situations that influence the ventilation process, ventilation methods and procedures, and tools and mechanized equipment used in ventilation. The manual includes chapter reviews, a glossary, and applicable safety considerations. 7th Edition (1994), addresses NFPA 1001.

FIRE SERVICE PRACTICES FOR VOLUNTEER AND SMALL COMMUNITY FIRE DEPARTMENTS

presents those training practices that are most consistent with the activities of smaller fire departments. Consideration is given to the limitations of small community fire department resources. Techniques for performing basic skills are explained, accompanied by detailed illustrations and photographs. 6th Edition (1984), 311 pages.

WATER SUPPLIES FOR FIRE PROTECTION

acquaints you with the principles, requirements, and standards used to provide water for fire fighting. Rural water supplies as well as fixed systems are discussed. Abundant photographs, illustrations, tables, and diagrams make this the most complete text available. It includes requirements for size and carrying capacity of mains, hydrant specifications, maintenance procedures conducted by the fire department, and relevant maps and record-keeping procedures. Review questions at the end of each chapter. 4th Edition (1988), 268 pages, addresses NFPA 1001, NFPA 1002, and NFPA 1031, levels I & II.

CURRICULUM PACKAGES

COMPANY OFFICER

A competency-based teaching package with 17 lessons as well as classroom and practical activities to teach the student the information and skills needed to qualify for the position of Company Officer. Corresponds to **Fire Department Company Officer**, 2nd Edition.

The Package includes the Company Officer Instructor's Guide (the how, what, and when to teach); the Student Guide (a workbook for group instruction); and 143 full-color overhead transparencies.

ESSENTIALS CURRICULUM PACKAGE

A competency-based teaching package with 19 chapters and 22 lessons as well as classroom and practical activities to teach the student the information and skills needed to qualify for the position of Fire Fighter I or II. Corresponds to **Essentials of Fire Fighting**, 3rd Edition.

The Package includes the Essentials Instructor's Guide (the how, what, and when to teach); the Student Guide (a workbook for group instruction); and 445 full-color overhead transparencies.

LEADERSHIP

A complete teaching package that assist the instructor in teaching leadership and motivational skills at the Company Officer level. Each lesson gives an outline of the subject matter to be covered, approximate time required to teach the material, specific learning objectives, and references for the instructor's preparation. Sources for suggested films and videotapes are included.

TRANSLATIONS

LO ESENCIAL EN EL COMBATE DE INCENDIOS

is a direct translation of **Essentials of Fire Fighting**, 2nd edition. Please contact your distributor or FPP for shipping charges to addresses outside U.S. and Canada. 444 pages.

PRACTICAS Y TEORIA PARA BOMBEROS

is a direct translation of **Fire Service Practices for Volunteer and Small Community Fire Departments**, 6th edition. Please contact your distributor or FPP for shipping charges to addresses outside U.S. and Canada. 347 pages.

OTHER ITEMS

TRAINING AIDS

Fire Protection Publications carries a complete line of videos, overhead transparencies, and slides. Call for a current catalog.

NEWSLETTER

The nationally acclaimed and award-winning newsletter, *Speaking of Fire*, is published quarterly and available to you free. Call today for your free subscription.

COMMENT SHEET

DATE _____ NAME _____

ADDRESS _____

ORGANIZATION REPRESENTED _____

CHAPTER TITLE _____ NUMBER _____

SECTION/PARAGRAPH/FIGURE _____ PAGE _____

1. Proposal (include proposed wording or identification of wording to be deleted),
    OR PROPOSED FIGURE:

2. Statement of Problem and Substantiation for Proposal:

RETURN TO: IFSTA Editor   SIGNATURE _____
            Fire Protection Publications
            Oklahoma State University
            Stillwater, OK 74078

Use this sheet to make any suggestions, recommendations, or comments. We need your input to make the manuals as up to date as possible. Your help is appreciated. Use additional pages if necessary.

COMMENT SHEET

DATE _____ NAME _____

ADDRESS _____

ORGANIZATION REPRESENTED _____

CHAPTER TITLE _____ NUMBER _____

SECTION/PARAGRAPH/FIGURE _____ PAGE _____

1. Proposal (include proposed wording or identification of wording to be deleted),
    OR PROPOSED FIGURE:

2. Statement of Problem and Substantiation for Proposal:

RETURN TO: IFSTA Editor   SIGNATURE _____
            Fire Protection Publications
            Oklahoma State University
            Stillwater, OK 74078

Use this sheet to make any suggestions, recommendations, or comments. We need your input to make the manuals as up to date as possible. Your help is appreciated. Use additional pages if necessary.